The Road to Maxwell's Demon
Conceptual Foundations of Statistical Mechanics

Time asymmetric phenomena are successfully predicted by statistical mechanics. Yet the foundations of this theory are surprisingly shaky. Its explanation for the ease of mixing milk with coffee is incomplete, and even implies that un-mixing them should be just as easy. In this book the authors develop a new conceptual foundation for statistical mechanics that addresses this difficulty. Explaining the notions of macrostates, probability, measurement, memory, and the arrow of time in statistical mechanics, they reach the startling conclusion that Maxwell's Demon, the famous *perpetuum mobile*, is consistent with the fundamental physical laws.

Mathematical treatments are avoided where possible, and instead the authors use novel diagrams to illustrate the text. This is a fascinating book for graduate students and researchers interested in the foundations and philosophy of physics.

MEIR HEMMO is an Associate Professor in the Department of Philosophy, University of Haifa. He has written on the foundations of quantum mechanics and statistical mechanics.

ORLY R. SHENKER is a Senior Lecturer at the Program for the History and Philosophy of Science, The Hebrew University of Jerusalem. She has written on the foundations of classical and quantum statistical mechanics and on the rationality of science.

The Road to Maxwell's Demon
Conceptual Foundations of Statistical Mechanics

MEIR HEMMO
University of Haifa

ORLY R. SHENKER
The Hebrew University of Jerusalem

CAMBRIDGE
UNIVERSITY PRESS

CAMBRIDGE
UNIVERSITY PRESS

University Printing House, Cambridge CB2 8BS, United Kingdom

Published in the United States of America by Cambridge University Press, New York

Cambridge University Press is part of the University of Cambridge.

It furthers the University's mission by disseminating knowledge in the pursuit of
education, learning and research at the highest international levels of excellence.

www.cambridge.org
Information on this title: www.cambridge.org/9781107424326

First published 2012
Reprinted 2013
First paperback edition 2014

A catalogue record for this publication is available from the British Library

Library of Congress Cataloguing in Publication data
Hemmo, Meir.
The road to Maxwell's demon / Meir Hemmo, Orly Shenker.
p. cm.
Includes bibliographical references.
ISBN 978-1-107-01968-3 (Hardback)
1. Maxwell's demon.   2. Second law of thermodynamics.
3. Statistical thermodynamics.   4. Maxwell, James Clerk, 1831–1879.
I. Shenker, Orly.   II. Title.
QC318.M35H46 2012
536'.71–dc23
2012012613

ISBN 978-1-107-01968-3 Hardback
ISBN 978-1-107-42432-6 Paperback

To my wife Tami, and to my children Alma, Avigail, and Shaul – M. H.
For my daughter Marie – may your trajectory, like that of
Maxwell's Demon, evolve through growing
potentialities and increasingly meaningful experiences – O. S.

Contents

Preface

This book is a product of more than decade of joint work during which we have greatly benefited from discussions with many people.

First and foremost, the approach we put forward here has been greatly influenced by David Albert's groundbreaking book from 2000 *Time and Chance*. Albert's way of thinking about Maxwell's Demon has made us realize that the foundations of statistical mechanics are in need of clarification, refinement and sometimes even revision, and this realization has led us to develop many of our ideas that come up in this book. Although we agree with Albert on some important matters – such as the radical idea that Maxwell's Demon is consistent with statistical mechanics – our approach substantially differs from his on a number of issues. However, the need to explain our differing opinions has greatly helped us sharpen our thoughts on the topics we address in this book.

Many conversations with the late Itamar Pitowsky over more than two decades have been extremely valuable to us. Itamar's open mind to new ideas and his constant encouragement throughout our research kept us on the right track.

It is a special pleasure to thank the members of the philosophy of physics group which meets monthly as it has for several years at the Edelstein Centre for the History and Philosophy of Science at the Hebrew University of Jerusalem: in particular we thank Yemima Ben-Menahem, Alon Drori, Daniel Rohrlich, Lev Vaidman, Boaz Tamir, and Simcha Rozen. We received valuable comments from the participants in two series of workshops in which we have presented some of our ideas: New Directions in the Foundations of Physics organized by Jeff Bub, Rob Rynasiewicz and James Mattingly, and the meetings in Sesto, Italy organized by GianCarlo Ghirardi, Nino Zanghi, Shelly Goldstein and Deltef Dürr; and from the participants in two international meetings of the Bar Hillel Colloquium for the History, Philosophy and Sociology of Science held at the Van Leer Institute in Jerusalem: in 2000, on the subject of foundations of statistical mechanics, and in 2008, on the subject of probability in physics, in honor of Itamar Pitowsky. We have also benefitted from discussions and written exchanges with (in alphabetic order) Frank Arntzenius, Joseph Berkovitz, Jeff Bub, Criag Callender, Shelly Goldstein, Amit Hagar, Carl Hoefer, the late Rolf Landauer, Barry Loewer, Tim Maudlin, Wayne Myrvold and Jos Uffink. We also thank the Cambridge University Press editorial team and especially Lindsay Nightingale for her thorough copy editing.

This research has been supported by the Israel Science Foundation, grant numbers 240/06 and 713/10, and by the German-Israel Foundation, grant number 1-1054-112.5/2009.

Meir Hemmo, Orly Shenker

1

Introduction

[Classical thermodynamics] is the only theory of universal content concerning which I am convinced that, within the framework of the applicability of its basic concepts, it will never be overthrown.[1]

In these words Albert Einstein expresses an unsolved tension in modern physics, which is the main topic of this book. A central aspect of our experience is that of an arrow of time. We feel a directedness from past to future and we encounter an asymmetry in time of many processes: it is easy to mix coffee with milk, but difficult to un-mix them; it is easy to break a glass or an egg, but hard to bring the pieces back together; people grow older, wood burns, petroleum becomes plastic – with no return. One might expect that these phenomena described by the laws of thermodynamics could be derived from the theories that govern the behavior of the particles that make up all matter. But, surprisingly, this is not the case. These time-asymmetric phenomena seem to have no root in the fundamental theories of physics. In particular, to any thermodynamic evolution that is possible according to the laws of classical mechanics, there corresponds a reversed anti-thermodynamic evolution that is equally possible. For this reason, classical mechanics cannot explain our experience of the thermodynamic asymmetry in time. It is widely believed that the gap between thermodynamics and mechanics can be closed by adding some probabilistic assumptions to mechanics, thus creating the theory of statistical mechanics. But this belief is based on several non-trivial assumptions that have not yet been proved, and there are good reasons to think that they will never be proved.

[1] Einstein (1970, p. 33)

1

This book provides a conceptual foundation for statistical mechanics. It shows how the gap between thermodynamics and mechanics comes about, and what is involved in the attempts to close it. The book ends with the conclusion that – contrary to Einstein's claim – thermodynamics is not a "theory of universal content" since it holds only for some special, albeit interesting, cases; and in other cases – especially those known as Maxwellian Demons – it can certainly "be overthrown." It turns out that the attempts to underwrite thermodynamics by statistical mechanics bring out issues that have implications for a variety of questions in physics and in philosophy. In this introduction we first present the problem of the gap between thermodynamics and mechanics, and then describe the general outline of the book.

Towards the end of the nineteenth century, two central theories in physics were put forward: thermodynamics and statistical mechanics. These two theories attempt to explain the same domain of phenomena but some of their predictions turn out to be incompatible with each other. According to the laws of thermodynamics, energy changes its form in such a way that it becomes less useful with time: a magnitude called entropy – which quantifies the amount of energy of an isolated system that cannot be exploited without investing work – increases in the course of time. The laws of thermodynamics were at first considered to be universally and invariably true of our world, without exception. However, it soon became clear that when the atomic structure of matter and the laws of mechanics that govern the motion of particles are taken into account, the thermodynamic laws must be understood *probabilistically*: Boltzmann, Gibbs[2] and other founders of statistical mechanics all agreed that these laws mean that the probability for entropy-decrease is extremely small, but not zero. It is this sort of probabilistic law that Einstein seems to think "will never be overthrown." Indeed, this has been the standard view, more or less, until today. According to that view, it is a consequence of the fundamental laws of statistical mechanics that the probabilistic version of thermodynamics is universally valid; and therefore anti-thermodynamic evolutions in which the entropy systematically and repeatedly decreases without investment of work are impossible.

In 1867, however, James Clerk Maxwell proposed a thought experiment that came to be known as Maxwell's Demon; he proposed this as a counter-example to the laws of thermodynamics in their absolute

---

[2] It is not clear whether Gibbs's approach to statistical mechanics is indeed genuinely probabilistic; see Chapter 11.

(non-probabilistic) version.[3] In Maxwell's thought experiment, "a finite being who knows the paths and velocities of all the molecules by simple inspection"[4] (which William Thomson (Lord Kelvin) called a Demon, although Maxwell insisted that it was a mechanical system[5]) takes a gas in a box at uniform temperature, with a partition in its middle, and manipulates the particles of the gas in such a way that a temperature difference is created between the two sides of the partition. The Demon observes the molecules of the gas, and then opens and closes the partition so that only the fast molecules pass to one side and only the slow ones to the other. After a while the fast molecules are separated from the slow ones, and this means that the temperature of the gas at one side is higher than at the other. The total energy of the gas is conserved, but its distribution changes, in such a way that it becomes more exploitable for the production of work – in clear violation of the laws of thermodynamics. Maxwell's thought experiment is based on the idea that mechanics is universal: there is no "framework of applicability" (to use Einstein's expression) in which the laws of thermodynamics reign and which is not subject to the laws of mechanics. The argument given by Maxwell's Demon implies that a serious adherence to the laws of mechanics entails that the laws of thermodynamics are not universally true.

Maxwell, however, also thought that the predictions of thermodynamics were good enough for many practical purposes: he thought that his Demon could not be practically constructed, for the sole reason that "we are not clever enough."[6] If we were clever enough, nothing in the fundamental laws of physics would stop us from imitating the Demon and acting in an anti-thermodynamic way. For this reason we take Maxwell to challenge also the standard view about the universal validity of the *probabilistic* version of the laws of thermodynamics.

For many years there have been numerous attempts at disproving Maxwell's view.[7] But in 2000 David Albert, in his book *Time and Chance*,[8] proved that Maxwell was right and that a Demon is consistent with classical mechanics. In this book we develop a conceptual framework for statistical mechanics that completes the project initiated by Maxwell

---

[3] See Chapter 13 for a detailed presentation of Maxwell's idea.
[4] See Knott (1911, pp. 213–214).   [5] See Knott (1911, pp. 214–215).
[6] See Knott (1911, pp. 213–214).
[7] See Leff and Rex (2003) for an extensive overview and annotated references.
[8] See Ch. 5 of Albert (2000). Albert assumes a view in philosophy of mind that is sometimes called a *physicalist* view (which we accept here) according to which there are no fundamental things or properties in the world that cannot be described by physics (see Chapter 5).

and Albert. Contrary to accepted wisdom, it turns out that a Maxwellian Demon is consistent with the principles of mechanics.

The route to understanding Maxwell's Demon will take us through a number of interesting ideas and open questions in modern physics. For example, we will need to examine the point at which the direction of time enters classical mechanics; the physical origin of macrostates; the meaning of probability in deterministic theories, how it arises, and in what sense it is objective and physical; how to give a physical account of observation, measurement, and memory in classical statistical mechanics; and finally what exactly is a Maxwellian Demon, and why it is consistent with classical statistical mechanics. And there are many other much more detailed questions we shall have to deal with on our way, even only partly, and even though we shall not have answers to all of them.

This book is not meant to replace standard textbooks in statistical mechanics. We focus on the conceptual foundations of classical statistical mechanics rather than on the physical application of the theory. In particular we focus on the question of what it means and what it would take to underwrite thermodynamics by statistical mechanics. These questions are addressed by standard physics textbooks, but some of the basic issues in these discussions are often unclear. For example, it is not always clear what macrostates are and how they come about; nor is it clear what exactly the meaning of probability is in statistical mechanics and how it arises given the deterministic nature of classical mechanics. These are the kinds of questions on which we focus, and we hope that the conceptual clarifications that we offer can then accompany the study of the further details found in standard physics textbooks on thermodynamics and statistical mechanics.

In our discussion we take the fundamental physical theory of the world to be classical mechanics. Of course, in contemporary physics classical mechanics has been replaced by quantum mechanics and the theory of relativity. Nevertheless, focusing on classical mechanics turns out to be fruitful, since the essential problems that arise in underwriting thermodynamics by the current quantum mechanical theory have already arisen in the context of classical mechanics and in almost the same way. And since quantum mechanics itself suffers from some serious interpretational issues, focusing on classical mechanics avoids unnecessary conceptual complexities. We consider some issues related to quantum mechanics in Appendix B.

Let us return to the above-mentioned problem in the attempts to underwrite thermodynamics by statistical mechanics, namely, the

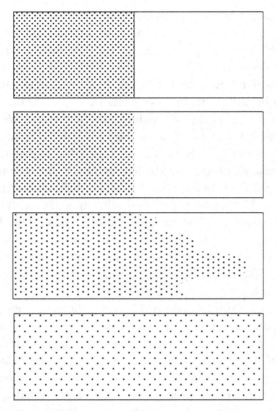

Figure 1.1 Gas expanding in a container

problem of the symmetry of evolutions in time.[9] Consider a film showing a gas expanding in a container; this is illustrated in Figure 1.1. We are all familiar with the plot of the film: in the beginning the gas is confined by a partition to the left-hand side of a container, and when the partition is removed the gas spreads out until, by the end of the film, it fills the entire volume of the container. We can easily tell, simply by looking at the film, whether it is shown in the order in which it was taken, or in the reversed order, i.e. from the end to the beginning. And the reason is that while we often experience processes similar to a gas that spreads out in a box, we never see anything like a gas that evolves spontaneously to a more concentrated state without some external intervention that exerts

---

[9] This is perhaps a misnomer: the problem is not about the direction of time itself, but rather about the direction of thermodynamic processes (and in particular of the direction in which entropy increases) in time.

considerable effort in order to compress it. The evolution in this film is characteristic of thermodynamic evolutions, and the asymmetry that is easily noticed in it is described by the time-asymmetric laws of thermodynamics.

Consider now an altogether different film, one which shows scenes taken from a billiard game. In these scenes we do not see the cues hitting the balls, nor do we see the balls falling into the pockets. We are shown only the parts of the game in which the balls bounce around, colliding with each other and with the rails around the table. Unlike the previous film, in this case we cannot tell whether the film is shown in the forward direction or in reverse. In this sense the billiard balls in this film exhibit a symmetrical evolution with respect to the direction of time. It is this time-symmetric behavior that is stated by the laws of classical mechanics.

The problem of the direction of time in statistical mechanics is the problem of how to reconcile these two apparently inconsistent pictures of the world. The problem is that the time-symmetric behavior of the billiard balls is taken to be characteristic also of the behavior of elementary particles, of which everything is made, including the gas in the first film, where time asymmetry prevails. Since the gas is made of particles, which behave in a time-symmetric way, we would have expected the gas as a whole to behave in a time-symmetric way. The fact that it does not calls for an explanation: how can the underlying symmetric mechanics give rise to the asymmetric phenomena in our experience? What assumptions beyond the time-symmetric laws need be added to mechanics in order to derive the observed thermodynamic phenomena, and on what grounds can these assumptions be justified?

Maxwell's Demon provides a way to understand these questions, since it can reverse the direction of thermodynamic processes: that is, it can effortlessly bring about a compression of the gas, thus reversing the direction in time in which the gas expands, simply by taking seriously the fact that the gas consists of particles that behave like billiard balls. In this book we describe the conceptual foundations of statistical mechanics, including everything that is needed from a physicalist point of view in order to understand the way in which Maxwell's Demon operates. In this sense, the book is a philosophical journey towards Maxwell's Demon. Here is a general outline of it.

In Chapter 2 we present quite briefly some of the main ideas in the theory of thermodynamics that are to be recovered by statistical mechanics. We present three main laws of thermodynamics: the Law of Conservation of Energy, and the two time-directed laws, the Law of Approach to

Equilibrium and the Second Law of Thermodynamics, which describe the thermodynamic asymmetric behavior in time. One main idea we attempt to convey in this chapter is that all the laws of thermodynamics are contingent empirical generalizations, and there is nothing *a priori* or conceptually necessary about them. For although this fact is never explicitly disputed, some arguments in the foundations of statistical mechanics – such as Einstein's words at the opening of this chapter – give the wrong impression that the status of the laws of thermodynamics is much stronger. It is this wrong impression that sometimes makes it hard to accept that Maxwellian Demons are possible. Another major idea we put forward in this chapter concerns the notion of thermodynamic entropy. We show that the physical content of the concept of entropy as characterizing the quality of different forms of energy – that is, as quantifying the degree to which energy can be exploited to produce work – holds only if the time-directed laws of thermodynamics are universally and unequivocally true (in their probabilistic version at least). In cases where the thermodynamic regularities do not hold, such as Maxwell's Demon, entropy no longer has this meaning.

In Chapter 3 we introduce the fundamental ideas of classical mechanics, and show that it is impossible to derive the time-directed laws of thermodynamics, in all their generality, from the laws of mechanics. We begin by introducing the central conceptual tools that will be used throughout the book, namely, microstates (which are the instantaneous states of the universe), the state space (which is the space of all the microstates a system can be in), the accessible region (which is the set of microstates that are compatible with the constraints on the system), macrostates (which are sets of microstates that are indistinguishable by observers, and can sometimes satisfy observable regularities), trajectories (which are the time evolutions of microstates), and dynamical blobs (which are bundles of trajectories all of which originate in some initial macrostate). Using these concepts, we outline the well-known no-go theorem, according to which the principles of mechanics contradict those of thermodynamics, unless the latter are weakened. We describe qualitatively and in outline central theorems in classical mechanics that will be used in later chapters, such as Liouville's theorem, Poincaré's recurrence theorem, and the ergodic theorem.

In Chapter 4 we consider the concepts of time and time-reversal invariance in classical mechanics. We argue that the direction or arrow of time, that is, the very distinction between past and future, is a primitive element of classical kinematics, which is already implicit in the definition of the microscopic mechanical state of a particle. Without such an arrow the most elementary mechanical state of a system is not well defined. Once this arrow is given, whether or not thermodynamic systems tend to approach equilibrium in the direction of time that we call from past to future is a matter contingent on the details of the equations of motion and on the initial microstate of the universe (where determinism entails that "initial" here may be any microstate along the trajectory of the universe). The thermodynamic behavior does not determine or define or even characterize the direction of time, which is a fundamental fact described by elementary mechanics. This direction is therefore fixed and given to us before the discussion of underwriting thermodynamics by mechanics even begins.

In Chapter 5 we construct the key notion in statistical mechanics, namely the notion of a macrostate. Macrostates – in our account – are sets of microstates that are indistinguishable by an observer, where this indistinguishability is an objective physical feature of the universe. The thermodynamic macrostates, which are the sets of microstates that appear as having the same thermodynamic properties, have a dual aspect which – we argue in this chapter – accounts for the thermodynamic phenomena. On the one hand, the thermodynamic macrostates are sets of microstates that are indistinguishable from one another by human observers; on the other hand, owing to the dynamics of the universe, these sets of microstates belong to sets of trajectories that satisfy certain dynamical regularities. Although these two features are both present in the thermodynamic macrostates, they are conceptually distinct and independent. Nevertheless, it is their combination that gives rise to the thermodynamic regularities: while the objective dynamical evolution, which determines the regularities, has nothing to do with our observation capabilities, the fact that we, human beings, happen to be able to distinguish between these macrostates explains the fact that we can perceive these regularities. It is the perception of the regularities which underlies the phenomena described by thermodynamics. There may be other sets of microstates that behave in a regular way, but since they do not match our perception, we are not aware of them, and we therefore have no theories about them.[10] We show in this chapter that Boltzmann's famous construction of macrostates, on the basis of his so-called combinatorial approach, is an example of macrostates that have this dual aspect. Since our account of

---

[10] Natural selection might explain this fit between the two aspects of the thermodynamic macrostates.

macrostates is based on the notion of indistinguishability, it obviously involves a notion of ignorance: an observer is ignorant about which of the microstates within a given macrostate is the actual microstate of the system. However, this indistinguishability is a consequence of the physical correlations between the observer and the observed system, and therefore we argue that this notion of ignorance is physically objective. Since macrostates are the key notion in statistical mechanics, and since they express the objective correlations between the observer and the observed, our construction of macrostates implies that the observer has an essential status in statistical mechanics. These ideas have implications not only for statistical mechanics but also for the philosophy of mind that we briefly address in this chapter: we show where exactly the mind–body problem enters the discussions of observers and measurement in physics. We believe that this discussion may be fruitful in the context of underwriting the so-called physicalist approach in philosophy of mind, since it addresses the gap between the microscopic physical structure of the world and the way we perceive it.

In Chapter 6 we address the question of how probability as an objective feature of the world arises in classical statistical mechanics on the basis of a completely deterministic dynamical structure. We argue that probability ought to be understood in this theory as transition probability between macrostates. In brief, this idea is as follows: the microstates that start in a given initial macrostate evolve so that their end points, at some given time, form a set that we call a dynamical blob. The probability that a system that starts in a given macrostate will end in some specified macrostate at a later specified time is given by the size of the overlap between that latter macrostate and the dynamical blob. The measure that determines the size of this overlap is chosen so as to fit the observed relative frequencies of macrostates in our past experience of similar evolutions. On the basis of this idea we argue that the size of a macrostate is completely irrelevant to the probability of this macrostate; this conclusion contradicts Boltzmann's view which is empirically inadequate since it inaccurately describes the approach to equilibrium. Our notion of probability suggests that observers who have more detailed observation histories may have an advantage in making accurate predictions, and we illustrate this idea in the case of the spin echo experiments. At the same time, we explain why observation histories do not matter in predicting thermodynamic behavior in normal circumstances. Finally, we address the prevalent understanding of statistical mechanics, based on the idea that the laws of thermodynamics can be underwritten by assuming some special probability distribution over the initial macrostate: although this approach is fundamentally mistaken, we describe in what cases statistical postulates concerning probability distributions over initial conditions can be made meaningful.

In Chapter 7 we apply the conceptual tools developed so far in order to construct the mechanical counterparts of the Law of Approach to Equilibrium and the Second Law of Thermodynamics: these laws are formulated in terms of the transition probabilities between macrostates. Since the notion of entropy is central to these laws, we describe its meaning in statistical mechanics. Thermodynamic entropy quantifies the degree of exploitability of energy, and therefore its proper mechanical counterpart is the degree of control of the system's microstate, and this degree can be expressed in terms of the size of the system's macrostate. The reason is that this size gives the average distance between the macrostate's boundaries, which we can manipulate to some degree, and the actual microstate. The measure which one should use to determine this size ought to be chosen on empirical data based on observations of thermodynamic evolutions. We stress that the measure of entropy is not necessarily the same as the measure of probability: these two measures have different origins, different empirical foundations, and altogether different meanings. Indeed, an important conclusion of our analysis of entropy in this chapter and of probability in the previous chapter is that, in general, entropy and probability are distinct and independent notions, and there is no dynamical or conceptual reason that implies that these two notions relate to each other in some particular way, so that for example macrostates of high entropy are more probable than macrostates of low entropy. On the basis of the notion of entropy, as well as the notion of equilibrium which we also present in this chapter, we outline what needs to be proven in order to establish the statistical mechanical counterparts of the laws of thermodynamics, and we describe an important attempt to do that, namely Lanford's theorem.

In Chapter 8 we distinguish between the formal notion of a probability measure and the notion of physical probability. Since there is an infinite number of probability measures that can be defined over a continuous set of points, such as the state space, we argue that a probability measure may be taken to correspond to physical probability only when the measure turns out to fit the relative frequencies of events in our experience. Therefore the probability measure chosen cannot explain our experience, as argued by the so-called typicality approach in statistical mechanics, but is rather explained by it. In particular, we argue that there are no natural measures in terms of which typicality claims can be formulated, and that, in particular, Liouville's theorem and the ergodic theorem which are often cited in this context are irrelevant for this choice. Finally, we consider the conditions under which it makes sense to introduce probabilities over initial conditions, and we show that in these conditions experience explains the probabilities, and not vice versa.

In Chapter 9 we focus on the question of what is a classical measurement. We argue that given its deterministic structure, classical micromechanics cannot account for a measurement interaction, and the understanding of this interaction is therefore at the level of macrostates. The dynamics of a measurement interaction is characterized by an evolution that we call split: in the split evolution the trajectories that start in the initial macrostate evolve into observationally distinct macrostates that correspond to the possible macroscopic measurement outcomes. The completion of a classical measurement necessarily involves what we call a classical collapse[11] of the dynamical blob onto the macrostate corresponding to the actual outcome of the measurement as perceived by the observer. We show how this idea is fully compatible with Liouville's theorem. We argue that, as opposed to the famous argument by Leó Szilárd (and also against the currently prevalent opinion based on the thermodynamics of computation), there is no intrinsic connection in statistical mechanics between measurement and entropy changes, and that the question of entropy change in measurement depends on the partition of the state space into macrostates. This partition is contingent on the details of the correlations between the microstates of the observer and the microstates of the observed system. In particular, we show that the total entropy of the universe may decrease during measurement. We draw conclusions concerning the essential status of the observer in statistical mechanics despite the observerless nature of the underlying classical mechanics.

Chapter 10 addresses the account that statistical mechanics can give of the past. According to our memories and records, which are summarized by the time-asymmetric laws of thermodynamics, entropy increases in the course of time, and was therefore lower in the past than it is now, and is likely to increase in the future. However, whenever the predictions of statistical mechanics are compatible with the laws of thermodynamics, retrodicting the past is inconsistent with our memories, since it implies

---

[11] This notion is radically different from the notoriously problematic notion of collapse in standard (von Neumann) quantum mechanics.

that entropy is highly likely to have been higher in the past than it is at the present, for any present time. This false inference is known as the problem of retrodiction or the minimum entropy problem. We analyze the standard solution to this problem known as the Past Hypothesis, according to which the retrodictions of statistical mechanics are taken to be unreliable and are replaced with inferences based on our memories, which are taken as reliable. We explain why, despite its appearance, this solution to the minimum problem is not ad hoc. In our view it turns out that memory and measurement have a parallel structure in the sense that both require similar conceptual resources, in particular the idea of a classical collapse onto the macrostate experienced by an observer. In measurement the collapse is onto the macrostate observed, and in memory it is onto the macrostate remembered by the agent. Finally, we show that the relation between the notion of memory and the notion of the past is not analytical: under quite general circumstances classical mechanics is compatible with trajectories along which we may remember the future just as we remember the past. Thus, whether what we remember is indeed in our past is contingent on the details of the segment of the trajectory along which we evolve.

In Chapter 11 we address the Gibbsian formulation of statistical mechanics. There are two major traditions in statistical mechanics, sometimes seen as two different theories, despite the same name and the same domain of application: one originates in the work of Gibbs and the other in the work of Boltzmann. The approach put forward in this book does not strictly belong to either of the two: we take from Boltzmann the notion of macrostates, which is missing in Gibbs's approach, but we give it a different meaning and a different role from the one it has in Boltzmann's approach; and we take inspiration from an idea by Gibbs in proposing the notion of a dynamical blob, which is missing in Boltzmann's (later) approach, but the meaning and role we give it are quite different from those of the original Gibbsian idea. In Chapters 5–10 we expressed Boltzmann's main ideas in terms of our conceptual framework, that is, in terms of the interplay between macrostates and dynamical blobs which gives rise to probabilities; and in this chapter we apply the same ideas to the Gibbsian approach. We provide ways to understand the use of probability measures, state averages, and functions of the measures in the so-called Gibbsian analogies of the thermodynamic quantities in terms of the conceptual framework of dynamical blobs, macrostates, and transition probabilities. We address the Gibbsian notion of thermodynamic equilibrium, the accounts of probability and of fluctuations in this approach, and the notorious coarse-graining argument which, for finite times, violates Liouville's theorem. The Gibbsian approach is the standard textbook version of statistical mechanics, and it yields empirically successful predictions. Nevertheless, some of the major ideas in the Gibbsian approach are considered by many to be ill-founded. This contrast between empirical success and shaky foundations calls for an explanation, and this is something we try to achieve in this chapter. Moreover, it turns out that the Gibbsian method (or our way of interpreting it) can provide us with a solution for a pragmatic problem in the application of our approach. By recasting Gibbs's approach in our terms we both explain its empirical success and provide it with a sound conceptual basis.

In Chapter 12 we address the notion of information, the physical implementation of logical operations and the thermodynamics of computation. According to the received opinion, first suggested by Rolf Landauer, logically irreversible operations such as erasure necessarily (that is, as a logical consequence of the principles of mechanics) involve entropy increase by $k \log 2$ per bit of erased information. We explain this thesis and put forward the necessary and sufficient condition for erasure (we call this condition *blending* dynamics) that is given in terms of the state space behavior of dynamical blobs and macrostates. We prove that while Landauer's thesis may hold in interesting cases, it does not logically follow from the principles of classical mechanics, and is therefore not universal and may be violated. This means that there is no special thermodynamic significance to the logical properties of the logical operations implemented in physical systems. The thermodynamic properties of such physical implementations depend only on the physical details of these processes. We further analyze some models of physically implementing logically reversible and irreversible operations and show precisely where the received opinion goes wrong.

In Chapter 13 we arrive at the final destination of our journey in this book, namely Maxwell's Demon. Applying the conceptual framework developed in the previous chapters we prove that a Maxwellian Demon is consistent with the principles of statistical mechanics. That is, a perpetual motion of a mechanical system, in which the total entropy of the universe systematically decreases, is possible. This, we believe, completes the project initiated by Maxwell and continued by Albert.

In Appendix A we give an example of a Demonic evolution on the basis of what is known in the literature as Szilárd's engine, which is a simplified version of Maxwell's original thought experiment. As is always the case, the concrete example involves numerous details, and brings into the discussion issues that are not directly related to the general argument in Chapter 13. Clearly, these issues need to be resolved before this particular example of a Demon becomes feasible. In this sense the concrete example may seem somewhat weaker than the general state space argument. However, we think it is still instructive to illustrate the way in which a Maxwellian Demon can be put to work, and to show how a thought experiment, similar in many ways to Maxwell's original one, should be analyzed in terms of macrostates and dynamical blobs. Perhaps more important, this example may shed some light on the sort of difficulties encountered on the way to its actual implementation, and this may help us understand why we are not surrounded by Demons despite their theoretical possibility and obvious benefits. As Maxwell wrote shortly after suggesting the Demon, the sole reason that we cannot construct Demons is that we are not clever enough. We do not – as yet – know how to overcome the obstacles, some of which are seen in our example. At the same time we emphasize that these obstacles are contingent on the specific details of this example, and do not at all weaken the general argument according to which a Maxwellian Demon is consistent with statistical mechanics.

In Appendix B we consider some issues related to quantum mechanics. It is sometimes said that since quantum mechanics is a probabilistic theory in which the probabilities are present already at the microscopic dynamical structure, the statistical mechanical macroscopic probabilities that feature in the recovery of the thermodynamic regularities could be entirely reduced to the quantum probabilities. In Appendix B.1 we address a proposal in this direction suggested by Albert, based on the spontaneous collapse theory proposed by Ghirardi, Rimini, and Weber, and we explain how questions concerning the explanatory role of typicality may arise also in this approach. In Appendix B.2 we consider the question of the meaning of the quantum mechanical probability distribution over initial conditions in Bohmian mechanics, and we question the explanatory role of typicality in Bohm's theory. And in Appendix B.3 we prove that a Maxwellian Demon is compatible not only with classical dynamics but also with quantum mechanical dynamics, both with and without collapse. We believe that this is enough to show that replacing classical mechanics with quantum mechanics as the basis for statistical mechanics does not counter Maxwell's intuition concerning the tension between thermodynamics and mechanics.

Finally, the literature on the conceptual foundations of statistical mechanics has become quite extensive in the past two decades or so. In what follows we refer only to the literature that is directly relevant to the issues we discuss, and we almost always avoid referring to primary sources. For an overview of various issues in this field and for substantial coverage of primary sources as well as secondary literature we refer the reader to Sklar (1993), von Plato (1994), Uffink (2007), Frigg (2008), North (2011), and Beisbart and Hartmann (2011).

2

Thermodynamics

2.1 The experience of asymmetry in time

A drop of ink mixes easily with water if gently stirred into it but cannot be un-mixed by stirring it in the opposite direction; ice melts, and boiled water cools down to room temperature, and then the water remains lukewarm indefinitely: it will never freeze nor boil spontaneously; smoke spreads in the air and never re-condenses. All these processes happen spontaneously and easily in one direction, but in order to reverse them complex processes have to be carried out and hard work has to be invested. Such *a-symmetric* processes surround us, and it is easy to come up with more examples. The asymmetry of these processes is *in time*: they are *time-asymmetric*, in the sense that we never encounter their reversed-order counterparts – that is, we never experience the reversals of processes such as those described above, in which the initial state and the final state exchange their order, such that the process traces its evolution backwards, returning to its initial state. All the time-asymmetric natural processes, like those described above, are manifestations of two general principles,[1] known as the *Law of Approach to Equilibrium* and the *Second Law of Thermodynamics*. The discovery of these general principles is one of the great achievements of science.

But what brings about this widespread experience of asymmetry in time? Is it a fundamental feature of the world, which we sense directly? Or is it a feature of the phenomena that emerge from some other, more fundamental occurrences? In that case – what is that fundamental state of

---

[1] Whether or not the Second Law of Thermodynamics is a law of nature depends on one's understanding of what laws of nature are. Since in this book we challenge the universality of the Second Law, it is appropriate to avoid giving it the title of a law of nature.

affairs which gives rise to the observed asymmetry in time? Or might it be that the time asymmetry is merely the result of the way that we experience the world, something that comes out of ourselves? In order to be able to phrase these questions in terms amenable to analysis in the framework of physics, we need to start with the theory that describes the asymmetry in time of our experience, namely, thermodynamics. In this chapter we will describe the theory of thermodynamics and show where the time asymmetry comes into play in it.

The theory of thermodynamics was developed during the second half of the nineteenth century. This was the century of the steam engine in locomotion, transportation, and industry; and because of the economic and social role of the steam engine, improving it had great importance. Improvement meant making stronger and faster engines that work for longer hours, using a smaller amount of coal. Quite surprisingly, the laws that govern steam engines turned out to be very general, and in fact as we shall see, in some important sense, they turned out to unify the whole of physics. These are the laws of thermodynamics.

The main laws of thermodynamics that we will describe here are the *Law of Conservation of Energy* (also called the *First Law* of Thermodynamics), the *Law of Approach to Equilibrium*, and the *Second Law* of Thermodynamics.[2] In the standard literature, the name "Second Law" is often given to the *combination* of what we call here the "Law of Approach to Equilibrium" and the "Second Law."[3] The distinction between these two laws was first made by Harvey Brown and Jos Uffink.[4] We believe that this distinction is vital to the understanding of thermodynamics. Indeed, as we will see, these two laws state different kinds of facts and require different kinds of explanations.

Roughly, the three laws just mentioned are interconnected as follows (a more detailed account is given in the following sections of this chapter). The Law of Conservation of Energy says that energy cannot be created *ex nihilo*, nor can it be lost. This law allows energy to change its form, for example from mechanical energy to heat energy, but its quantity is conserved. However, when the form of energy is altered, something happens to it: the degree of its exploitability may change. The degree to which energy is exploitable in order to produce work depends on the form

---

[2] For a presentation of the First Law and the Second Law of Thermodynamics, as well as the rest of standard thermodynamics, see Fermi (1936).
[3] For the meaning of the *Second Law* in different contexts see Uffink (2001).
[4] Brown and Uffink (2001).

in which this energy is stored in a system. In general, it is an experimental fact that energy tends to change its form in such a way that the amount of energy that is exploitable as work decreases with time. Some of the changes in the forms of energy and in its distribution tend to happen spontaneously, i.e. without our having to intervene in the process; this fact is described by the Law of Approach to Equilibrium and the Second Law. Consider, for example, the energy of the molecules in a waterfall. As is well known, when the molecules are at the top of the fall their potential mechanical energy can be easily exploited to produce useful work. However, this same quantity of energy is useless when it is transformed into kinetic energy in the disordered flow of water molecules at the bottom of the fall. The Law of Conservation of Energy is satisfied but, nevertheless, something happens to the energy: the degree of its exploitability decreases. This practical quality of energy is called its *entropy*, and so the empirical generalization in the Law of Approach to Equilibrium and the Second Law is that entropy (on the whole and in general) increases; by historical convention, low entropy means high exploitability. And so, the combination of the three laws – the Law of Conservation of Energy, the Law of Approach to Equilibrium, and the Second Law of Thermodynamics – tells us that the *quantity* of energy in the universe is constant, but its *quality* tends to decrease with time. The thermodynamic arrow of time is therefore the arrow of the decrease in the quality of energy in the universe.

We now turn to describe these laws of thermodynamics in some detail. A thorough understanding of these laws is extremely important for our purposes in this book, namely to challenge the thermodynamic arrow of time, that is, the idea that the decrease in the quality of energy in the universe is a universal law of nature. This challenge is the idea behind the thought experiment known as Maxwell's Demon, and here we begin our journey towards this aim.

2.2 The Law of Conservation of Energy

Before considering the notion of energy in thermodynamics, we start with energy in classical mechanics.[5] According to Newton's Second Law of mechanics (not to be confused with the Second Law of Thermodynamics), if a force F acts on a particle, the particle's velocity changes in accordance with the law of motion $F = ma$, where a is the particle's acceleration

---

[5] A more detailed account of mechanics is given in Chapter 3.

(change of velocity with time) and m is the particle's mass. A useful way to describe the effect of a force is to describe the change that this force brings about in the state of the particle, that is, the change it brings about in its position and velocity. But this description omits information concerning the force itself, whose action depends on properties of the particle as well as its environment. A description taking these factors into account is given by a state function known as *energy*. A *state function* is a function of the system's position and velocity and of its internal properties in terms that are relevant to indicate the action of the forces.[6] Therefore, through the change of energy we can learn about the change not only of the state of the system, but also about the way in which this change has been brought about by the action of forces. For example, if a ball with mass m is released and falls freely from height h above the Earth, the force of the Earth's gravity transforms the initial potential energy of the ball mgh (where g is a constant) to kinetic energy $\frac{1}{2}mv^2$ (where v is the velocity of the ball), and this change in the form of energy both describes and accounts for the ball's final velocity.

This last example illustrates the contents of the Law of Conservation of Energy, which states that the energy state function is such that while its form may change, its total quantity in the universe remains constant. In the above example, $mgh + \frac{1}{2}mv^2 = $ constant. What is the basis for this law? As long as the only force is gravity, the conservation of energy as expressed by $mgh + \frac{1}{2}mv^2 = $ constant is a *theorem* of Newtonian mechanics.[7] But if other forces are involved and other theories come into play, the law expresses an extremely powerful and exceptionless generalization of our *experience*. One beautiful illustration of this idea – which has also played a part in shaping this law[8] – is Joule's experiment, in which mechanical energy $(mgh + \frac{1}{2}mv^2)$ is transformed into *thermal energy*, called *heat*. Let us consider Joule's experiment in some detail.

The experiment, performed by James Prescott Joule around 1845, is described schematically in Figure 2.1. Our description of Joule's way of

---

[6] Energy is nothing but a function of the state of matter. In this sense energy is not part of the classical ontology, and its status is different from the status of matter. In the special theory of relativity the relationship $E = mc^2$ shows that matter and energy are interchangeable, but nonetheless this does not mean that they are identical. In any case, our discussion here is classical and not relativistic.

[7] In mechanics, the Law of Conservation of Energy can be understood as expressing some basic symmetry. See Brading and Castellani (2003) for discussions and references.

[8] Here we do not give an historical account of the discovery of the Law of Conservation of Energy and of the precise role of Joule's experiment in this discovery; such an account is given by, for example, Harman (1982).

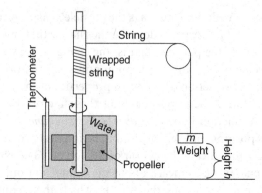

Figure 2.1 Joule's experiment

thinking is anachronistic: we do not attempt to recover it but rather to illustrate the way in which the significance of his experiment is understood today. At the initial stage of the experiment, at time t_0, a weight of mass m is prepared at some height h above some surface. The potential energy of the weight in this initial state is mgh. We now release the weight and let it fall to height zero, such that its potential energy becomes zero. According to the mechanical theorem of conservation of mechanical energy, this energy cannot be lost and, in this case, it is transformed into kinetic energy $\frac{1}{2}mv^2$ of the weight. This kinetic energy of the weight is transformed into kinetic energy in the propeller, which is connected to the weight by a string, as shown in the figure, and as a result the propeller starts to revolve in the water. However, as is known from experience, the friction between the propeller and the water slows down the propeller until it stops revolving and its kinetic energy becomes zero.

At first glance, this process seems to suggest that energy is not conserved, but can be lost. However, Joule's great discovery was that when the propeller stopped moving the temperature of the water was higher than its initial temperature. The idea of conservation of energy is then saved by conjecturing that *heat is a form of energy*: if it is, then the rise in the temperature of the water may correspond to a rise in the internal energy of the water by the exact amount of the mechanical energy that prima facie seemed to have been lost. If that is correct, then energy can change its form between gravitational potential energy, kinetic energy, and heat energy, and only their sum must be conserved: $mgh + \frac{1}{2}mv^2 + \text{heat} = \text{constant}$. The formal expression of the Law of Conservation of Energy is $\Delta E = Q - W$ where E is the total energy, Q is the heat energy that enters the system (taken, conventionally, to be positive), and W is the mechanical energy produced (conventionally regarded as positive). Obviously nothing in Joule's experiment *proves* that heat energy and mechanical energy are two forms of the same thing. However, if we assume that ultimately thermodynamics should be underwritten by mechanics, Joule's identification of heat with energy seems justified. This is important for subsequent chapters in which we will make a connection between thermodynamics and mechanics.

The notion of heat, although central to thermodynamics, is not a simple one, and it is worth considering further its theoretical status. As Joule's experiment illustrates, changes in internal energy can be measured by changes of temperature. If we see that the temperature has changed, we assume that the internal energy has changed. But the ratio between temperature and heat differs between different substances, and so we need to calculate the amount of heat in a different way, not via temperature. If the source of energy is wholly mechanistic, this is simple: we calculate the loss of mechanical energy elsewhere and conclude that this quantity has been transformed into heat. But what if the source of energy is heat from another system, such as heat that is transferred to water from fire? How can we calculate the amount of this heat? The way to do it is via the Law of Conservation of Energy, as follows. We imagine a purely mechanical process which would bring about the same increase in temperature in the same substance, and calculate the amount of mechanical energy that would be invested in the system in that case, and then, applying the Law of Conservation of Energy, we conclude that the amount of heat energy that has been given to the system is exactly the same as in the imaginary mechanical process. The quantity of heat in the real experiment is taken to be the same as the quantity of mechanical energy in the imaginary experiment, because both lead to the same final state. From this point of view, in thermodynamics, heat is a *residual* concept that we can calculate only if we assume the Law of Conservation of Energy. In order to account quantitatively for heat in a non-residual way we shall need to turn to statistical mechanics.

The Law of Conservation of Energy may be thought of as a unifying principle of the whole of physics. The idea of conservation of energy has been generalized to other forms of energy, beyond mechanical energy and heat, using reasoning of the kind we will now demonstrate on the basis of Joule's experiment. As we know, we can raise the temperature of water by, say, burning coal. Since the initial and final states of the water are (by assumption) the same no matter how we raise their temperature (that is,

whether we do it by the propeller or by burning the coal), we can infer that
the amount of heat energy transferred from the coal to the water is exactly
equal to the amount of mechanical energy given to the water by the propeller. Generalizing the Law of Conservation of Energy we can then conclude
that this amount of energy was initially stored in the coal in the form of
chemical energy – another form of energy that takes part in the energy
transformations. A similar reasoning holds with respect to other forms of
energy such as electrical energy. In this way, Joule's and other experiments
can be taken as demonstrating that all forms of energy are interchangeable
(subject to the constraints expressed by the other laws of thermodynamics
discussed below). This move means that the Law of Conservation of Energy
unifies different theories of physics by identifying in all of them magnitudes
having the proper units, such that the sum of all these magnitudes taken
together is conserved. Of course, while the search for such magnitudes may
be motivated by the desire to satisfy the Law of Conservation of Energy, the
successful identification of these magnitudes and the empirical verification
of the conservation of their sum mean that the principle is not a tautology
but has empirical significance.[9] This unifying role gives the Law of Conservation of Energy its special importance in physics.

2.3 The Law of Approach to Equilibrium

Joule's experiment, which demonstrates the interchangeability of various
forms of energy, also illustrates the following puzzling phenomenon.
Mechanical forms of energy are fully interchangeable: for example, as a
ball bounces off a floor, the gravitational potential energy that the ball
has at the top of its motion is fully transformed into kinetic energy at the
floor level, and this kinetic energy is fully transformed back into gravitational potential energy as the ball rises back into the air.[10] Now, in Joule's
experiment, all the mechanical energy stored in the weight is transformed

---

[9] The First Law is sometimes treated as if it were analytic in some sense. For example, in
1775, before the full formulation of the First Law of Thermodynamics (the Law of
Conservation of Energy), the French Academy of Sciences (L'Academie Royale des
Sciences) declared that it would no longer consider research papers proposing devices
that violate this law. To see that by this declaration the First Law has been given a status
above and beyond that of empirical generalization, and assumed a status not far from an
a priori truth, it is instructive to note that at the same declaration the Academy declared
that it would no longer consider certain *other* kinds of proposals either. These were
papers purporting to square the circle, trisect the angle, or duplicate the cube.

[10] Of course, we assume here the ideal case where the collision with the floor is elastic.

easily into heat energy: all we have to do is release the weight and let it fall, and the rest happens *spontaneously*, all by itself, until we are left with the weight on the floor and the water heated up. This process has a clear direction, and it never occurs in the reverse order: once the propeller has been stopped by the friction with the water, and the heat energy has spread in the water, the system reaches a stable state, in which it remains indefinitely (unless an external agent intervenes[11]). The process seems to have a natural *direction*, towards a predictable stable state of *equilibrium* (a notion we discuss below). This directionality in time is described by the Law of Approach to Equilibrium which describes the thermodynamic arrow of time, and it is this law that we will challenge on our journey to Maxwell's Demon. Therefore we now turn to discuss it in some detail.

The Law of Approach to Equilibrium is not a theorem of thermodynamics, but rather a generalization of our experience. Therefore, the best way to introduce it is by illustrating its content using an example, which is paradigmatic in thermodynamics, owing to its obvious importance for steam engines. Consider the example illustrated in Figure 1.1 in which a gas is confined by a partition to the left-hand side of a container, filling half of the container's volume, and then the partition is removed. Once the partition is removed, nothing hinders the gas from expanding in the container. In other words, suppose that the gas is no longer confined to the left half of the container, and the only restriction on its evolution – the only *constraint* on it – remains the volume of the whole container. We know from experience that the gas will evolve spontaneously in such a way that after some relatively short time it will uniformly fill up the entire volume of the container, and that once it reaches this state the gas will remain in it indefinitely.[12] This is a special case of the Law of Approach to Equilibrium, which says that given a set of constraints, there is a unique final state, called equilibrium, towards which a thermodynamic system evolves spontaneously, which it reaches after some finite time (not asymptotically), and in which it remains indefinitely. In Part 1 of Figure 1.1 the gas is constrained by the partition, and the corresponding state of equilibrium is the one in which the gas fills up the left half of the container; and in Parts 2, 3, and 4 of the figure the constraint is the entire volume of the container, and the corresponding equilibrium state is the one in which the gas fills up this volume.

---

[11] We study this case later on in the context of the Second Law of Thermodynamics.
[12] Note that the relaxation time after which the system arrives in equilibrium depends on the properties of the substances involved.

The Law of Approach to Equilibrium is schematic in the sense that it does not specify which state is the equilibrium state for any given set of constraints. This fact needs to be learned from experience: for example, the fact that the particular equilibrium state of a gas in a container is the one described in Part 4 of Figure 1.1 is not a consequence of the Law of Approach to Equilibrium, but is something we know from experience with gases in containers. An important example of an equilibrium state is a state of uniform temperature. Take two gases at different temperatures, which are brought into thermal contact. It is a fact known from experience that heat flows from the hotter gas to the colder one until the temperature of the combined system becomes uniform. Temperature can be thought of as the tendency of a system to give away heat. And as the tendencies of the two systems equalize, heat no longer flows from one to the other. In this evolution the total energy of the gas remains constant, but its distribution evolves towards equilibrium. The state of uniform temperature is the state of equilibrium, and therefore given an initial state in which the temperature is not uniform one can predict the direction of the thermodynamic evolution, from the Law of Approach to Equilibrium. To sum up: this law says that there is a direction towards which thermodynamic systems evolve in time, but it does not specify the details of the end state or the time it takes for the evolution to it.

The Law of Approach to Equilibrium also says that when the system arrives in the state of equilibrium that is determined by the external constraints, it remains there indefinitely. In fact, the stability of the equilibrium state is part of the very notion of equilibrium in thermodynamics, which means that when a system is in such a state, all its thermodynamic properties (such as volume, pressure, and temperature) are constant over time. It is, however, instructive to realize that there is something rather circular in this definition of equilibrium, in the sense that thermodynamic magnitudes are *defined* only for states of equilibrium and are undefined otherwise. Of course, we can *see* systems in out-of-equilibrium states: for example, if the gas in Figure 1.1 is colored and expands slowly enough, we can see it in various stages of its expansion, before it reaches the equilibrium state in which it fills up the entire container. But the magnitude of volume in the theory of thermodynamics is defined as the volume of the gas when it reaches equilibrium. This definition is not unreasonable, since the standard measuring devices (such as thermometers and barometers) give well-defined and repeatable results of thermodynamic magnitudes only in equilibrium states. However, the theory is clearly *incomplete* in this sense. In order to discuss magnitudes in out-of-equilibrium states we will need statistical mechanics.

The evolution described by the Law of Approach to Equilibrium is asymmetric in time, since it is invariably headed towards equilibrium rather than away from equilibrium. The Law of Approach to Equilibrium expresses a thermodynamic arrow of time, in the sense that if we are given a series of states of a thermodynamic system we can easily determine their order by using this law, since the closer a state is to equilibrium, the later it is in time, relative to the other states, according to this law.[13] Whether or not this arrow of time is a universal consequence of the fundamental laws of physics is a question we examine later on.

2.4 The Second Law of Thermodynamics

Joule's experiment illustrates the fact that mechanical energy can be transformed into heat energy; that is, it illustrates the Law of Conservation of Energy. Moreover, it illustrates the Law of Approach of Equilibrium since by the end of the process the set up halts in a state of equilibrium, much like our example of a gas spreading in a container. We now want to focus on a third aspect of Joule's experiment, namely the asymmetry of energy transformations between different equilibrium states, which illustrates the Second Law of Thermodynamics. There are actually two equilibrium states in Joule's experiment: before the experiment begins, the system is in some sort of equilibrium where the water is at a low temperature, the propeller is steady, and the weight is held in place, say, on some shelf; call this equilibrium state *A*. Then the shelf is removed and the experiment begins and ends in the equilibrium state described above; call this equilibrium state *B*. Can we reverse the temporal order of these states and devise another experiment in which before it begins the system is in equilibrium state *B*, and after it ends the system is in equilibrium state *A*? Of course, the new experiment cannot be a simple reversal of Joule's experiment, since the Law of Approach to Equilibrium would prevent such a process; but can there be another, different experiment, one that will bring about this result? The answer is no, and this

---

[13] This arrow of time need not be understood as suggesting some sort of teleological evolution; it does *not* say that the equilibrium state is a target state rather than a consequence of some causal evolution. Indeed, if the fundamental ontology of the universe conforms to physics, then we should accept a causal account. We will see later on that the explanation of the laws of thermodynamics on the basis of classical mechanics is patently causal rather than target-oriented.

negative answer is given by the Second Law of Thermodynamics: this law prevents heat energy from being fully transformed into mechanical energy, as we will now explain.

The Second Law is, then, another thermodynamic arrow of time: while the arrow of time in the Law of Approach to Equilibrium consists of ordering states in time according to their distance *from equilibrium*, the arrow of time in the Second Law of Thermodynamics orders states *of equilibrium* in time, according to which the amount of energy which is in the form of heat, relative to the amount of energy which is in mechanical or other forms of energy (which are more readily exploitable to produce useful work), cannot decrease. In other words, the Second Law describes an *arrow of time* in terms of entropy, namely the degree of exploitability of energy in order to produce work. When we proceed to discuss statistical mechanics, we will show that the two thermodynamic arrows of time are closely connected, and in an important sense they are a single arrow; but as long as our discussion is confined to the theory of thermodynamics, we have here two *distinct* arrows of time.

In order to present the Second Law of Thermodynamics we will describe another case study of historical importance: the Carnot cycle. The Carnot cycle is an abstraction of the operation of a heat engine which is, in turn, a great simplification of a steam engine, which was, as we have said, the topic of research that led to the creation of thermodynamics. A heat engine is a system intended to transform energy in the form of heat into mechanical energy, and is therefore directly relevant to the problem presented above, regarding the entropic arrow of time.[14] Carnot's cycle reveals and illustrates the fundamental limitations on energy expressed by the Second Law of Thermodynamics.

The Carnot cycle operates in the following way, illustrated by Figure 2.2. We prepare a gas in a container subject to several constraints, as illustrated by Stage 1 in the figure. The first constraint is the temperature and internal energy of the gas, and is imposed by coupling the gas to an environment at temperature T_H, where H stands for "high." At the beginning of Stage 1 the gas is in equilibrium with this environment. When two systems in thermal contact arrive at equilibrium their temperatures are equal. Normally, their

---

[14] Carnot worked in the context of caloric theory of heat, and so our description of his argument as pertaining to energy in the form of heat is anachronistic. The aim of our discussion is not historical, but rather to explain the significance of the Carnot cycle to the way in which thermodynamics is understood today.

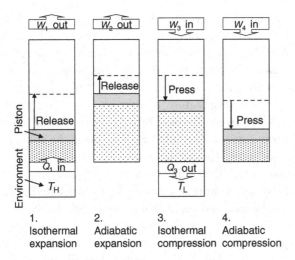

Figure 2.2 The Carnot cycle

final common temperature is somewhere between the initial temperatures of the two systems; but here we make the customary assumption – to be examined later – that since the environment is vast the heat it transfers to the gas is relatively negligible and has no effect on the temperature of the environment, and therefore when the gas and its environment reach thermal equilibrium, the temperature of the gas changes but the temperature of the vast environment remains constant. The result is that the gas arrives at the temperature of the environment, T_H. (This customary assumption is not trivial, and we will question its status in the context of statistical mechanics later in the book.)[15]

The second constraint on the gas at the beginning of Stage 1 of Figure 2.2 is a force F exerted on the piston by an external agent, which determines the pressure p of the gas. Since by assumption at the beginning of Stage 1 the whole system is in equilibrium, the volume of the gas at that moment is constant, and this indicates that the force exerted by the pressure p of the gas on the piston is exactly equal (and in opposite direction) to the force F of the external agent.

It is convenient to start by assuming that the gas constrained by the environment and by the piston is an *ideal gas* – although later the Carnot

---

[15] Thermodynamics is silent about the nature of matter. In particular it does not even insist that matter is made of particles, and in fact as long as one works within thermodynamics (and does not proceed to statistical mechanics) one may assume, for all practical purposes, that matter is continuous. In this sense thermodynamics is autonomous from mechanics.

cycle is generalized to the case of other sorts of gases. An ideal gas is an idealization of a diluted gas obeying the so-called *ideal gas law* (sometimes known as Boyle's Law after its discoverer), according to which whenever the gas is in a state of *equilibrium*, its pressure and volume are directly proportional to its temperature: $pv = NRT$ (where p is the pressure, v is volume, T is temperature, N is the amount of gas given in moles, and R is a constant). In equilibrium, by definition, p, v and T are constant over time. This means that the volume of the gas in the container is fixed by our two constraints, T_H and the pressure p.

At the beginning of Stage 1 the whole system is in equilibrium, subject to the above two constraints: the environment that determines the temperature T_H of the gas, and the external force equal to the pressure p of the gas. If the external constraints were to remain constant the system would remain in this equilibrium forever. In order to activate the heat engine, the equilibrium has to be disrupted, by a change of the external constraints; this starts a *thermodynamic causal chain*, as follows:

The external agent weakens the force exerted on the piston from F to $F - \Delta F$, thereby releasing the piston gradually and very slowly; ideally the change is infinitesimally small, in which case the process is called *quasistatic* (we expand on this term below). Whenever the piston is released by some amount, the following chain of processes is generated, owing to the Law of Approach to Equilibrium. First, since the pressure p of the gas on the piston is met by a weaker force $F - \Delta F$, the gas pushes the piston and extends in volume to $v + \Delta v$. The result is a reduction in the pressure of the gas, and this continues until the force due to new pressure $p - \Delta p$ becomes equal to the external force $F - \Delta F$, at which point the gas reaches a new state of pressure equilibrium with the external agent that presses the piston. This expansion of the gas from volume v to volume $v + \Delta v$ means that the gas exerts the force of its pressure on the piston along a distance Δx, and this means that the gas is doing *mechanical work* on the piston. Work is a mechanical notion that connects the notion of energy with that of force by giving the amount of energy transferred by a system that exerts the force. Work is given by $W = F\Delta X$, the action of a force along a distance; in cases such as the Carnot cycle where the force comes from pressure and the distance is given by an increase of volume, this expression translates to $W = \int p dv$.[16] Since the gas does work on the piston, it

---

[16] The idea behind this translation is this. Since pressure is the force acting on a unit of area, we have: $[p][v] = \frac{[F][x^3]}{[x^2]} = [F][x]$ (where the square brackets denote units).

transfers an amount of energy equal to $\int p dv = W_1$ to the external agent. According to the Law of Conservation of Energy, the internal energy of the gas decreases by the same amount. Since temperature is determined by the internal energy, the decrease in internal energy means that the temperature of the gas decreases from T_H to $T_H - \Delta T$. But this leads to a temperature difference between the gas and its environment, and then, according to the Law of Approach to Equilibrium, heat Q_1 in an amount equal to W_1 flows from the hot environment to the gas until the temperature of the gas returns to T_H. Such a process is called *isothermal*. As we said, in thermodynamics it is customary to assume that the environment is so large that the transfer of heat Q_1 to the gas is, in effect, negligible in the sense that it does not change the temperature of the environment.

The above process continues: the external agent reduces the force exerted on the piston, and consequently the gas *expands isothermally*, until the gas reaches a pre-designated volume. This means that the external agent merely releases the piston while the hot environment pushes the piston. The net result is that the external agent can use this to produce work equal to W_1. The paradigmatic example is of course a steam engine.

Notice that this description of Stage 1 is clearly time-directed, from the beginning of Stage 1 to its end. This time-directedness is a result of the Law of Approach to Equilibrium (so that in order to understand it we are not yet in need of any additional law; in particular we do not need the Second Law of Thermodynamics that will be presented later on). According to the above description, the process begins by creating a very small deviation from the equilibrium, brought about by the external agent which releases the piston, and this deviation starts a chain of processes in which the system evolves to equilibrium in such a way that one small deviation from equilibrium is followed by another, until everything is back to equilibrium: the *non*-equilibrium of pressure, brought about by the decrease in the external force, results in an *approach* to equilibrium by way of expansion but this, in turn, brings about a *non*-equilibrium of temperature, which results in an *approach* to equilibrium by way of heat entering the gas from the environment. But this description disagrees with a customary idealization of the process, according to which the process is *quasi-static*, which means the following. It is customary to assume that, since all the deviations from equilibrium described above are *infinitesimal*, one can say that, to a very good approximation, *the gas is in equilibrium throughout the process*. Of course, taking this idealization too seriously would mean that the system is never away from equilibrium and therefore never evolves towards equilibrium, in which case the thermodynamic causal chain does not work. And so the idealization is that despite the fact that deviations from equilibrium are the driving force of the whole process, whenever we observe the system we will find it in equilibrium. Strictly speaking this idealization is incoherent. Although for all practical purposes if the process is slow enough the deviations from equilibrium are undetectable, their existence is central to the very possibility of the process, and therefore they are *not negligible*.[17]

By the end of Stage 1 we have transformed heat energy Q_1 taken from the environment into work W_1 given to the external agent. We now wish to exploit *more* heat from the environment in order to produce *more* work. Since letting the gas expand indefinitely is not practical, the way to achieve this aim is to bring the gas back to its initial state at the beginning of Stage 1, and then let it start again. This is called closing the thermodynamic cycle, and is described in Stages 2, 3, and 4 of the Carnot cycle. As we will shortly see, it turns out that closing the cycle is highly non-trivial and is subject to constraints imposed by the Second Law of Thermodynamics.

At Stage 2 of Figure 2.2 we disconnect the thermal contact between the gas and the environment, and we let the gas expand further, as the external agent continues to release the piston by reducing the force exerted on it, very slowly (ideally quasi-statically). A process such as this carried out when the gas is thermally insulated is called *adiabatic*. During this stage of *adiabatic expansion* the thermodynamic causal chain described above is repeated except that no heat flows into the gas. As the gas expands and pushes the gradually released piston, it produces work equal to $\int p dv = W_2$, thereby transferring energy in the amount of W_2 to the external agent. Conservation of energy implies that the internal energy of the gas decreases by this amount, and since no energy flows into the gas from its environment, its temperature decreases. We let the process continue until the temperature of the gas reaches a designated temperature T_L which is lower than the original temperature T_H.

At Stage 3 we again couple the gas to an environment, but this time to an environment which has the low temperature T_L. The external agent now *increases* the force exerted on the piston, again very slowly (ideally quasi-statically), thereby compressing the gas in the container. As a result, again, a series of small deviations from equilibrium and subsequent approaches to equilibrium occurs. Since this time the external agent

---

[17] See Sklar (2000) on the acceptability of idealizations in physics.

invests work in the gas, the internal energy of the gas increases, and consequently its temperature increases from T_L to $T_L + \Delta T$. At this point the Law of Approach to Equilibrium entails that heat must flow out from the gas to the cold environment, until the gas reaches temperature T_L. This chain is repeated until, by the end of the *isothermal compression* of Stage 3, we have transformed mechanical energy taken from the external agent into heat given out to the cold environment. (We would have liked to avoid such a stage which means a waste of work, but as we will see shortly this stage is unavoidable.) We continue this process until the gas reaches some designated volume.

At Stage 4 we disconnect the thermal contact between the gas and the environment, and continue to compress it adiabatically until its temperature reaches T_H. By the end of this stage the gas is returned into its initial thermodynamic state, as we planned to do by the end of Stage 1, and the cycle of operation can start again. The net balance of exchanges of work and heat implied by conservation of energy is:

$$Q_1 = \sum_{i=1}^{4} W_i + Q_3.$$

This means that some of the heat extracted from the hot environment is wasted and not turned into work.[18]

The Carnot cycle is reversible in the sense that it can operate in the reversed order: the order of stages is reversed, and within each stage the direction of the evolution is reversed; see Figure 2.3. Let us examine the meaning of this more closely. In the reversed cycle, a thermodynamic causal chain, of small deviations from equilibrium and small evolutions towards equilibrium, is activated in the reversed direction, by letting the external agent push the piston at the stages where it previously released the piston, and releasing the piston at the stages where it previously pushed it. As a result the directions of the heat flow in the reversed cycle are reversed.

We start with Stage 4 (of Figure 2.3) at which the gas is at T_H and the external agent releases the piston extracting work until the gas cools down to T_L. This is a heat pump: the external agent cools down the gas without investing work. But again, we need to close the cycle in order to continue

---

[18] These conclusions are derived from calculations of the work $W = \int p dv$ produced and invested along the cycle; see Fermi (1936).

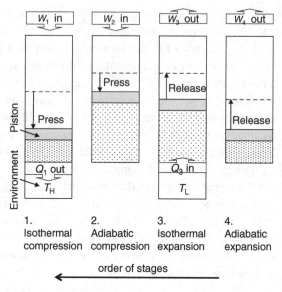

Figure 2.3 The reversed Carnot cycle

the process, and this is why we need to go through the subsequent stages in the reversed order, 3, 2, and 1. At Stage 3 the gas, now coupled to the cold environment, expands isothermally. At Stage 2 the gas, now insulated, is compressed adiabatically, and the compression continues isothermally at Stage 1 until the cycle is closed. The net balance of work and heat is given, as before, by the above equation, but now the net result is that work is invested and heat is transferred from the cold environment to the hot one. The difference between the two directions of the cycle is that in the forward direction some of the heat taken from the hot environment is transformed into work and the rest to the cold environment; while in the reversed direction some of the invested work is transformed into heat in the hot environment and the rest is received back, thus pumping heat from the cold environment. In the reversed direction the gas traces exactly the same equilibrium states, in the reversed order. But the reversed cycle differs from the forward cycle in the details of the small processes in which the gas approaches equilibrium, that is, in the non-negligible deviations from equilibrium and approaches to equilibrium, which drive the cycle.

Carnot proved that despite the apparent inefficiency of the cycle his engine is the most efficient one possible: for any given temperatures of the environments, the amount of work produced by the Carnot cycle in the forward direction is the highest and the amount of work invested in the reversed direction is the lowest. Carnot's proof is a consequence of the Second Law of Thermodynamics.[19]

This description of the net result of the cycles is based on an important idealization made throughout the argument, namely, that the environments have an infinite heat capacity in the sense that the heat that flows between the gas and the environments leads to no change in the temperatures of the environments. That is, there is no intrinsic connection between the heat energy in the environment and its temperature. This is a bizarre idea since it rejects from the outset that temperature and energy are related in the way brought out by Joule's experiment. The idea that the Carnot engine operates in a cycle rests entirely on this idealization. Given this idealization, the Carnot cycle can be repeated indefinitely and transform heat energy to work, as long as there is some energy left in the hot environment. If we relax this idealization, we obtain that after each run of the cycle the temperature difference between the two environments decreases and therefore the gas does not in fact return to its initial state. When the two environments reach the same temperature the engine will no longer work. From this perspective the difference in temperature between the two environments is itself a state of non-equilibrium and the whole system evolves, one cycle after another, towards uniform temperatures; and when this thermal equilibrium is reached the engine halts, in accordance with the Law of Approach to Equilibrium.

Let us now focus on an interesting feature of this cycle, which will lead us to the Second Law of Thermodynamics. Recall the question we raised with respect to Joule's experiment: can there be a process (different from the one in Joule's experiment) which will take the set up from the end equilibrium state in which the weight is low and the water is warm to an equilibrium state in which the weight is high and the water is cool? It might seem that Carnot's cycle could be such a process, since it carries out the desired transformation of energy, from heat (in the hot environment) to mechanical energy (given to the external agent). However, the Carnot cycle seems to carry out this transformation in a fundamentally *inefficient* way, since not all the heat energy which it takes from the hot environment is transformed to mechanical energy: as we saw, in every cycle of operation some energy is lost to the cold environment. And so a Carnot cycle

---

[19] See Fermi (1936).

that would extract heat from the warm water in Joule's experiment would fail to take Joule's set up back to its initial state, since it would not transform all the heat into work and therefore the weight would not return to its original height. But why use such an imperfect engine? Can we not construct a better engine that will transform *all* the heat absorbed during Stage 1 of the cycle from the hot environment into a form of energy that is useful to do work, and so prevent the heat being lost to the cold environment?

The answer is no, and the generalization of this answer is the Second Law of Thermodynamics. And the significance of this answer is that not only is energy unavoidably wasted in all processes, but also that this loss constitutes an arrow of time: the arrow that says that equilibrium states in which energy is less exploitable as work occur later than equilibrium states in which more energy is exploitable as work.

The Second Law has been formulated in two canonical forms which are provably equivalent, and which are generalizations of what we saw in the forward and in the reversed operation of the Carnot cycle. In the forward direction of the cycle we saw that some of the heat absorbed by the gas from the hot environment is lost to the cold environment. This fact was generalized by Lord Kelvin and elevated to the level of a physical law as follows:

> A transformation whose only final result is to transform into work heat extracted from a source which is at the same temperature throughout is impossible.[20]

In the reversed direction of the cycle we saw that in order to transform heat from a cold environment to a hotter one, work has to be invested. Clausius generalized this fact by the following law:

> A transformation whose only final result is to transfer heat from a body at a given temperature to a body at a higher temperature is impossible.[21]

These two formulations of the Second Law are provably equivalent.[22] To get a feeling of the proof in terms of the Carnot cycle, consider the following. Suppose that we carry out the Carnot cycle in the forward direction and, as we saw, we lose some heat to the cold environment. Now suppose that contrary to Clausius's formulation we can take this amount of heat and bring it back to the hot environment without investing work.

---

[20] Fermi (1936), p. 30.   [21] Fermi (1936), p. 30.
[22] See Fermi (1936, Ch. III), Albert (2000, Ch. 2 and Appendix).

If this were true, then the net amount of heat that actually left the hot environment would have been entirely transformed into work, in violation of Kelvin's formulation. Suppose now that contrary to Kelvin's formulation we transform all the heat absorbed from a single environment to work without any other change. By friction we can transform all this work into heat, and heat up any system we choose including a system that is hotter than our environment. This is a violation of Clausius's formulation. And so, violating either of the formulations of the Second Law leads to a violation of the other, and therefore they are equivalent.

The Second Law of Thermodynamics (in both formulations) implies that energy which is not already in a purely mechanical form is never fully exploitable to produce work. Some energy is always lost, typically by becoming heat through some sort of friction or through heat flow of the kind we saw in the Carnot cycle. More generally, the energy of a system can be divided into two parts: a part that is exploitable as work and a part that is not. This idea is expressed by the equation

$$\Delta E = \int p \mathrm{d}V - \int T \mathrm{d}S,$$

where ΔE is some change in the internal energy of a system between two equilibrium states, $\int p \mathrm{d}V$ is the work produced during that transformation,[23] and $\int T \mathrm{d}S$, where T is the temperature of the system, is taken to be the change of energy to a form that is not exploitable as work. The quantity ΔS is called the *entropy difference* during the energy transformation. This is in essence the Second Law.

In the Carnot cycle, for example, ΔE is the energy in the form of heat Q_1 given to the gas at Stage 1, $\int p \mathrm{d}V$ is the net amount of work that is given to the external agent during a full cycle, and $\int T \mathrm{d}S$ is the amount of heat Q_3 lost to the environment at the low temperature T_l at Stage 3.

But what is precisely the physical content of the quantity ΔS? Clausius suggested that ΔS is calculated by integrating the quantity $\delta Q/T$ along the evolution,[24] where Q is the heat generated or absorbed during an infinitesimal time interval and T is the temperature during that interval. However, it is crucial to see that this quantity expresses the degree in which the

---

[23] Here one may add other components pertaining to other ways of exploiting energy to produce work, for example by chemical potentials.

[24] Here δ expresses the mathematical notion of imperfect differential. The precise significance of this fact is not relevant for our present discussion.

energy in a system is exploitable to produce work *only in processes which satisfy the Second Law*, namely only if the Second Law of Thermodynamics is true for these processes. In evolutions which violate the Second Law, such as Maxwell's Demon, the quantity $\delta Q/T$ does not correspond to exploitability of energy. In other words, the statement that $\delta Q/T$ is associated with exploitability of energy is not a mere definition, but a statement about the world, and the content of this statement is that the Second Law of Thermodynamics is *true* of the world. If the Second Law is *not* true, then the integral over $\delta Q/T$ has nothing to do with the degree to which energy is exploitable. This is somewhat striking, since in the usual way of thinking about thermodynamics one automatically thinks about the concept of entropy as expressing exploitability of energy. But this is wrong, and the origin of this mistake is the received wisdom that the Second Law of Thermodynamics expresses an absolute and universal truth about the world. It is this wisdom that we challenge by proving that Maxwellian Demons are consistent with classical mechanics.

Since ΔS is a difference between two equilibrium states, one can define the entropy of an equilibrium state relative to some standard equilibrium state, and thus get

$$\Delta S = \int_A^B \frac{\delta Q}{T} =:_{\text{def}} S(B) - S(A).$$

If A is the standard state one can define its entropy to be 0, and this difference will yield $S(B)$. On the basis of the formulations of the Second Law by Kelvin or Clausius, one can prove that the entropy difference

$\Delta S = \int_A^B \frac{\delta Q}{T} =:_{\text{def}} S(B) - S(A)$ between *any* two equilibrium states A and B

of an isolated system (such as the whole universe, or the Carnot cycle *including* its two environments), along *any* path starting in equilibrium state A and ending at equilibrium state B, satisfies the following inequality:

$$S(B) \geq S(A),$$

where the equality holds for reversible transformations. (Recall that this expression, like all expressions in thermodynamics, holds for equilibrium states only, since the thermodynamic magnitudes are defined only for equilibrium states.) This latter inequality is known as the *Second Law of Thermodynamics in terms of entropy*, and it says that in an isolated system entropy cannot decrease.[25] Since the entropy formulation assumes those of Kelvin and Clausius, and the formulations of the Second Law by Kelvin and Clausius, in turn, can be derived from the entropy formulation of the entropy law, all three formulations are *equivalent*.

To get a feeling for the connection between exploitability of energy and entropy difference (where this connection holds, that is, where the formulations of Kelvin and Clausius are *true*), consider the expression for the entropy difference of an ideal gas between two equilibrium states A and B: $\Delta S = c_V \ln(T_B/T_A) + R\ln(V_B/V_A)$, where c_V denotes the molecular heat capacity at constant volume, and R is a constant. We say that the entropy in state B is higher than the entropy in state A since the energy is less exploitable in state B than in state A, and this can happen in, for example, the following two situations. The first is expressed by the term $\ln(V_B/V_A)$: suppose that $V_B > V_A$. Then, at state A we can exploit energy by letting the gas expand to the larger volume in state B; this ability is lost once we reach the state B. The second case is expressed by the term $\ln(T_B/T_A)$: suppose that the temperature of the environment is T_B and the temperature of the gas is T_A, and that $T_B > T_A$; then we can build a heat engine between these two temperatures. But, once the temperatures of the gas and the environment become equal, this is no longer possible.

We can now apply the Second Law in order to solve the puzzle of the apparent inefficiency of the Carnot cycle. Recall that the puzzle is that the other laws we discussed – the Law of Conservation of Energy and the Law of Approach to Equilibrium – do not, by themselves, hinder a transformation of *all* the heat absorbed from the hot environment to work; and so one might think that it could be possible to construct a more efficient engine that – unlike the Carnot cycle – will not lose energy to the cold environment. For instance, one might think that it would be possible to construct a system that will take the end state of Joule's experiment, in which the weight is low and water is warm, and transform the system back to the pre-experiment equilibrium state in which the weight was high and the water was cool. However, it turns out that the Second Law of Thermodynamics implies that this is impossible: if one assumes the Second Law (in any of its forms), one can prove that the Carnot cycle is the most efficient engine, in the following sense. Given two environments in given temperatures T_H and T_L, and given some substance as the gas on which the engine works, the Carnot cycle transforms the

---

[25] For a discussion of various formulations of the Second Law, see Uffink (2001).

maximal amount of energy from the hot environment to work, and loses the *minimal* amount of heat to the cold environment. And so *the Second Law of Thermodynamics implies that one cannot do better than the Carnot cycle* in transforming heat energy to useful work.

2.5 The status of the laws of thermodynamics

The laws of thermodynamics, described above, are generalizations of robust experience. However, these generalizations are robust only in the appropriate circumstances for which they were meant to apply, and they are empirically violated outside their intended domain of application. Most conspicuously, observable phenomena exhibiting fluctuations are abundant: the temperature of systems is never absolutely constant, and neither are its pressure and volume. It is an empirical finding that accurate measurements reveal fluctuations. The state of systems that may initially appear to be already in equilibrium is not steady: it exhibits small fluctuations away from their so-called equilibrium state and back; larger fluctuations are less frequent than small ones, but still, fluctuations are the rule.

Fluctuations violate a strict formulation of the Law of Approach to Equilibrium as well as of the Second Law, both of which have no room for exceptions or fluctuations. In view of these phenomena it is customary not to take thermodynamics too seriously,[26] and to weaken its laws, i.e. to replace them by *statistical counterparts*, in order to preserve at least some of their original ideas. The behavior, which is strictly forbidden by the original laws of thermodynamics, is described in these counterparts as merely extremely improbable.

This change is often taken to be a small and pragmatically acceptable correction but, in fact, the change is profound. Since thermodynamic magnitudes are defined only for systems in equilibrium, standard thermodynamics has no conceptual tools to describe fluctuations. Replacing the strict laws of thermodynamics with probabilistic counterparts requires a profound conceptual change in this theory, which includes a change in the way in which thermodynamic magnitudes are defined and measured. Instead, the usual way of dealing with the probabilistic turn involves leaving thermodynamics and entering the domain of statistical mechanics.

---

[26] See Callender (2001).

This step, however, entails a radical revision of the status of the laws of thermodynamics: they become effects of the mechanical behavior of the constituents of the thermodynamic system.[27] One of the strong consequences of moving from thermodynamics to statistical mechanics is that it opens the way to Maxwell's Demon, that is, a mechanical system which is a counter-example to the Second Law and to the Law of Approach to Equilibrium. However, the Demon violates not only the laws of thermodynamics, but also their intended *statistical* mechanical counterparts, as we show in this book.

The situation is, then, that there are two sorts of threats to the universal validity of the Second Law of Thermodynamics and the Law of Approach to Equilibrium. The first threat is empirical, namely the phenomenon of fluctuations. This threat is met by replacing the absolute laws with probabilistic counterparts. The replacement is not carried out within thermodynamics, but rather by leaving standard thermodynamics and entering statistical mechanics. This step, however, gives rise to the second threat, namely that of Maxwell's Demon, to the universal validity of the laws of thermodynamics – but this time the threat is to the validity of the *probabilistic* counterparts of these laws. In other words, Maxwell's Demon is a threat to the possibility of proving even the probabilistic versions of the time-directed laws of thermodynamics from the principles of mechanics.

---

[27] This is an expression of a reductionist view, which we indeed accept but will not argue for in this book. The kind of reduction we have in mind will become clear as we proceed.

3

Classical mechanics

3.1 The fundamental theory of the world

By the end of the previous chapter we introduced two threats to the universal validity of the laws of thermodynamics. The first threat is empirical, and is the phenomenon of fluctuations. This phenomenon requires replacing the strict laws of thermodynamics by probabilistic counterparts, according to which behavior such as evolution away from equilibrium is not strictly forbidden but merely highly improbable. Since standard thermodynamics has no conceptual tools that can express this idea, a transition to probabilistic laws means a transition from thermodynamics to a different theory. A natural candidate for replacing thermodynamics in order to describe the phenomenon of fluctuations is statistical mechanics. In this field, fluctuations are easy to describe and account for, and probabilistic counterparts of the Law of Approach to Equilibrium and the Second Law of Thermodynamics can be formulated. And, at this point, the second threat to the universal validity of thermodynamics emerges: the theory of statistical mechanics allows for cases, known as Maxwellian Demons, which contradict even the probabilistic counterparts of the Law of Approach to Equilibrium and the Second Law of Thermodynamics. Maxwell's Demon is a thought experiment, based on the assumption that the world is, fundamentally, mechanical. In Chapter 13 we will present the idea of Maxwell's Demon, but to understand this thought experiment and its significance it is necessary to start by describing the theory of mechanics, on which it is based.

We start out, then, by taking classical Newtonian mechanics to be the fundamental theory of the world. Of course, taking classical Newtonian mechanics as the fundamental theory of the world can be only a *working hypothesis* and not a serious claim, for, as is well known, classical mechanics has been replaced by quantum mechanics and by the special and general theories of relativity; and it is fair to say that according to current physics, classical Newtonian mechanics is believed to be *false*.[1] Our decision to take classical mechanics as our working hypothesis may seem puzzling: what is the point, one might wonder, in judging the adequacy and scope of thermodynamics on the basis of a theory which we think is false? The reason is that the main problems encountered in classical statistical mechanics are repeated – with rather small conceptual changes – in the transition to quantum statistical mechanics; but the presentation of these problems is much simpler in the classical case. Of course, the problems in the mechanical theory *itself* are quite different as one moves from classical mechanics to quantum mechanics; but the problems associated with proceeding from mechanics to statistical mechanics are very similar, whether one moves from *classical* mechanics to classical statistical mechanics, or whether one proceeds from *quantum* mechanics to quantum statistical mechanics.[2] And so it seems to us preferable to carry out the discussion in the context of the theory, which simplifies the presentation of the ideas that we wish to convey.

And so, in this book, we set out to examine the relationship between classical statistical mechanics and thermodynamics. The central problem concerning this relationship had already been recognized by Maxwell as early as 1867, when the theory of statistical mechanics was just beginning to crystallize. Maxwell had made his point by way of his thought experiment of the Demon. The blunt and simple point of the Demon argument is that if classical mechanics is the fundamental theory of the world, and if there is no flaw in the Demon argument, then the Second Law of Thermodynamics and the Law of Approach to Equilibrium are not universally true, even in their weak probabilistic versions. In this book we set out to show that the Demon argument is valid. As we said, we are going to present and illustrate our arguments using classical mechanics. It is therefore necessary to begin by first understanding the basic features of classical mechanics which are central to the foundations of statistical mechanics and the underwriting of

---

[1] Of course, classical mechanics is pragmatically useful and can yield approximately correct predictions in the appropriate circumstances.
[2] We show this in a number of appendices. Moreover, it is important to study classical statistical mechanics since in many cases one gets the impression that problems in quantum mechanics are solved merely by reducing the quantum behavior to a classical one, implicitly assuming that in the classical case most of the central problems have been already solved. In this book we show that this is not the case.

thermodynamics by mechanics, and which play a role in the Demon argument. This is the aim of the present chapter.

3.2 Introducing classical mechanics

It is a fundamental premise in classical mechanics[3] that the world is made up of particles.[4] Richard Feynman expressed this idea as follows.

> If, in some cataclysm, all of scientific knowledge were to be destroyed, and only one sentence passed on to the next generation of creatures, what statement would contain the most information in the fewest words? I believe it is the atomic hypothesis that all things are made of atoms.[5]

The laws governing the behavior of atoms, or more generally of particles, are – by our working hypothesis – the laws of classical mechanics.[6] Since thermodynamic systems, such as the systems in Joule's experiment and in Carnot's cycle, are nothing but collections of particles, they must obey the laws that particles obey – namely, the laws of mechanics. In other words, the fact that thermodynamic systems are made of particles that obey the laws of mechanics ought to underwrite the laws of thermodynamics, to the extent that these laws are true.[7]

---

[3] What exactly is the content of the theory of classical mechanics, and what are its auxiliary hypotheses, is an open question, with which we do not deal here; for us, in this book, the theory includes the ontology as we present it here. For a discussion of the scope of the theory see, for example, Hutchison (1993) and Callender (1995).

[4] Classical physics introduces fields, in addition to particles. These include the electric and magnetic fields. Fields are extended in space and in this sense differ significantly from particles. The microscopic state of the world at a given time actually consists of the microstates of the particles and the microstates of the fields. Presumably, in the above quotation Feynman understands that fields are produced by particles; we do not take a stand on this point. In this book we do not discuss fields explicitly, since we are interested in the distinction between microstates and macrostates, and our distinction can be extended to a discussion of fields. The *microstate* of a field at a given time is the complete and precise description of the field at this time, and the fact that fields are extended in space is perfectly compatible with their having microstates in this sense. A *macrostate* of a field is a *set* of counterfactual microstates of the field compatible with some observable properties (see Chapter 5). Since the problems in statistical mechanics that arise from the ontology of particles arise just as well in the context of fields, we prefer to talk about particles only.

[5] Feynman (1963, vol. 1 pp. 1–2).

[6] When talking about laws of nature we may be understood as referring to the regularities in nature; we make no metaphysical point regarding the nature of laws.

[7] Again, we take here a reductionist approach in the sense specified in the text, which is standard in the foundations of statistical mechanics. For ideas concerning emergent properties, see Butterfield and Bouatta (2011) and Butterfield (2011).

The collections of particles that make up thermodynamic systems are enormous (for example, one litre of gas consists of about 10^{23} particles), and so even if they are *in principle* subject to the laws of mechanics, the application of these laws in order to account for the behavior of such collections and to predict their evolution is impossible: one cannot solve, and in fact one cannot even write down equations of motion for such collections. However, there is a way out of this difficulty, which is described by Lev D. Landau and E. M. Lifshitz in the following words.

> At first sight we might conclude. . . that, as the number of particles increases, so also must the complexity and intricacy of the properties of the mechanical system, and that no trace of regularity can be found in the behavior of a macroscopic body. This is not so, however, . . . when the number of particles is very large, new types of regularity appear.[8]

The hope is that statistical mechanics would carry out the project proposed by Landau and Lifshitz, and express the new kind of regularity that governs the behavior of bodies consisting of large numbers of particles. In order to see how this can be done we shall first describe some general features of the laws of mechanics.

3.3 Mechanical states

When we say that mechanics governs the behavior of particles we mean to say that the laws of mechanics allow us to describe the evolution of *states of particles*. Given the state of a particle at some point of time, and given some other information about the particle and about its environment, we can predict the state of the particle at other times, using Newton's famous Second Law of Mechanics, $F = ma$. The notion of state here is well defined: the mechanical state of a particle at a point of time is its position and velocity.[9] It is this notion of state that mechanics can predict. And if we wish to underwrite the laws of thermodynamics by the laws of mechanics, we need to show that the mechanical states of particles evolve in a way that satisfies the laws of thermodynamics. To do so we must begin by describing the way in which mechanical states are represented in mechanics. (Notice that the notion of a state is theory dependent, and that other

---

[8] Landau and Lifshitz (1980, p. 1). Although this may sound as though it is a statement about emergence, it is not.

[9] More precisely, the mechanical state is given by the generalized position and momentum. We stick to the position and velocity for simplicity of presentation.

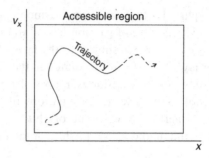

Figure 3.1 State space with accessible region and a trajectory segment

theories describe states differently. For instance, the thermodynamic state of a system is well defined only for equilibrium, and is characterized by magnitudes such as volume, pressure, and temperature.)

The mechanical state of a single particle at a given point of time is described by six numbers: three numbers give the particle's position in the three directions of space, and three give the corresponding directions of velocity. For reasons that will become clear below, these numbers are called *degrees of freedom*.[10] It is convenient (for the purposes of our discussion) to describe the state of a particle as a point in a six-dimensional space, called the *state space* or *phase space*. In Figure 3.1 we depict two representative dimensions out of the six dimensions of the state space of a particle: the position along the x direction of space and the velocity v_x in the x direction.[11]

Particles are never free to have any state whatsoever out of the infinity of states in their state space. They are always subject to limitations on the velocities that they can have, and often on the positions they can be in. The limitations on the possible velocities a particle can have are brought about by the fact that particles have finite amounts of energy, and limitations on possible positions can be brought about by, for example, confining a particle to a container. These limitations determine a region in the state space that contains all the states a system can have, which are compatible with those limitations. This region is called the *accessible region* (see Figure 3.1). As we will see in later chapters, the notion of accessible region is key to understanding statistical mechanics as an

---

[10] If the particle is subject to certain kinds of limitations or constraints, or if there are correlations between the degrees of freedom of the particles, their number is smaller, as we explain below.

[11] Alternatively one can think of the figure as depicting a two-dimensional projection of a six-dimensional space.

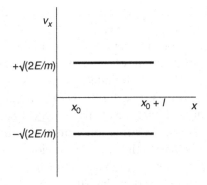

Figure 3.2 The accessible region of a particle with energy E on a string of length l

objective physical theory, and for this reason we expand further on this notion.

To illustrate the notion of accessible region in the state space, consider the simple case of one particle that is confined to move along a string, which is stretched in the x direction and has length l. Suppose that the particle has internal energy E and that no external forces (such as gravitation) act on it, so that all its energy is kinetic (see Figure 3.2). The accessible region – that is, the region that contains all the positions and velocities that the particle can have – is determined by these limitations, as follows. First, since all the energy of the particle is kinetic, that is, $E = mv_x^2/2$, the only possible velocities that this particle can have are $v_x = \pm\sqrt{2E/m}$, either to the right or to the left along the x direction. This is a rather limited degree of freedom but, technically speaking, it is still a degree of freedom in the sense that the particle can have either of these two velocities, given its total energy; the total energy does not determine *which* of these two velocities the particle has at a given time, but only that no other velocities are possible. The position degree of freedom is slightly more flexible: the constraint on the particle is consistent with finding the particle in any position in the l interval along the x direction. These two degrees of freedom determine that the accessible region of the particle consists of two line segments in the (x,v_x) plane, as illustrated in Figure 3.2. The end points of these two line segments are determined by the appropriate values of x (say, some x_0 and $x_0 + l$) and $v_x = \pm\sqrt{2E/m}$. In addition, these points are determined by the (not depicted) positions of the ends of the string in the y and z directions, and by the fact that the velocities in the y and z directions are $v_y = v_z = 0$. Each end point of each line segment is defined by six numbers but, in the present case, only two of them are actually degrees of freedom in the sense that only the values of x and v_x can have more than one value.

The question remains, of course, what determines *which* of these values the particle has at each point of time: which of the two velocities $v_x = \pm\sqrt{2E/m}$ and which of the infinitely many positions in the interval $[x_0, x_0 + l]$ the particle will assume during its evolution. It is important to notice that nothing in the limitations imposed on the particle, by its total energy and confinement to the string, prevents the particle from being in any of the states that belong to the accessible region which consists of the two line segments in the figure, at any point of time. The answer to the question of which state the particle has at any point of time is given in terms of the particle's *dynamics* and its *initial conditions*, which we discuss below.

This example shows that there are two types of constraints that determine the accessible region. The first type determines the *number of degrees of freedom*: a particle on a string has two degrees of freedom, out of the six numbers that determine its state; the other four are not "free" in the sense that they have only one possible value. The second type of constraints reduces the *range along each of the degrees of freedom*. In the case of the particle on a string, the length of the string limits the range of the positions of the particle to the interval $[x_0, x_0 + l]$, and the energy limits the available velocities in the x direction to two values only.[12]

This case can be easily generalized to the case of a particle in a container. Here, the limitation on the positions available to the particle is a straightforward generalization of the string example: in the case of the container the particle is free to move within ranges corresponding to the dimensions of the container (along the spatial directions x, y, and z). The

---

[12] In the example above, of the particle on a string, the intuitive background assumption is that the particle is the system of interest while the string is the external constraint defining the accessible region. This intuitive approach goes back to Newton who focuses on the particles and the forces that act on them. Although this intuition is useful in giving a physical interpretation to the mathematical formalism of the theory, it may also be misleading. As we saw above, in order to understand the physics of the particle on a string, for instance, to see what the possible physical states of the particle are, we must focus on the set of degrees of freedom of the system. In fact, in mechanics the system is nothing but these degrees of freedom (see Lanczos 1970, pp. 3–4). This point, once appreciated, has an important consequence: the identity of the system is defined by the sum total of the constraints to which it is subjected and the parameters associated with it. Once the constraints are taken into account, the accessible region is well defined and the system behaves in this region freely. That is, the behavior of the system within the accessible region cannot be controlled *by construction*, and its behavior is determined entirely by the equations of motion (discussed below). For simplicity, and wherever there is no danger of confusion, we shall refer to a system as particles subject to constraints.

total energy determines the speed of the particle, that is, the vectorial sum of the three components of velocity, in the three directions. At each point of time the speed of the particle is given by $|v| = |\sqrt{2E/m}|$, but since the direction of motion can change, the projections of this speed on the different directions can vary. For this reason we have a range of possible velocities in each direction, and the range is the continuous interval $[-\sqrt{2E/m}, +\sqrt{2E/m}]$ along any of the three directions. In fact, owing to the limitation $|v| = |\sqrt{2E/m}|$, given the velocity of the particle in any two directions (say x and y) one can calculate the velocity in the third direction (z, in this case). Since that third component of the velocity is determined by the energy together with the other two components of velocity, the velocity of the particle in that direction is actually not free. In that sense we say that the energy reduces the total number of degrees of freedom by one. Using a geometrical analogy, reducing the dimensionality of a three-dimensional space by one means having a two-dimensional surface; and therefore we say that the total energy of a system determines an *energy hypersurface* to which the state of the system is confined. In the event that the energy is the only constraint on the particle, the entire energy hypersurface is the accessible region. Additional constraints confine the particle to subregions of this hypersurface.

Since our project here is to underwrite the state of thermodynamic systems by mechanics (or rather to see whether such underwriting is possible), we need to focus on such systems (such as the Carnot cycle), which are characteristically composed of many gas particles in a container. Let us begin with representing two particles in a container, and then generalize to the case of many particles. One way to represent the state of two particles is within the state space of a single particle: here, the state of two particles is represented by two points. This method is known as the *μ-space* method.[13] The μ-space is the state space of a single free particle (or molecule, hence the letter μ), which has six dimensions, standing for the three spatial directions and three directions of velocity of a single particle. However, the description of the two-particle state as two points in the μ-space is wanting, since it cannot account for constraints imposed on the state of each particle owing to correlations between these states. For instance, think of two particles connected by a rod of length l: although each particle can be anywhere in the accessible region, given that *one* particle is at some position a, the other can only be

---

[13] See Ehrenfest and Ehrenfest (1912).

in a position on the surface of a sphere of radius l centered on a. This means that the correlation between the two particles imposes limitations on the two-particle state, which cannot be expressed in the μ-space. In order to account for such correlations we represent the particles in the state space of two particles (or – equivalently – phase space or Γ-space). The state space of two particles has $2 \times 6 = 12$ dimensions, six for each particle. In this space each point represents the combined state of the two particles. In general, the state space of a system consisting of N particles is $6N$-dimensional, and the mechanical state of a system is represented in this space by a single point. In such a space, the total energy of the N-particles system determines an energy hypersurface of $6N - 1$ dimensions, to which the state of the system is confined. Normally, there are other constraints on systems as well, limiting the number of degrees of freedom and the ranges of values along each degree of freedom, and all these limitations together make up the system's accessible region.

It is instructive to compare the role of constraints in mechanics and the role of constraints in thermodynamics, for it seems that they are slightly different. Recall that in thermodynamics, the Law of Approach to Equilibrium says that the external constraints determine the equilibrium state uniquely: to any set of constraints there corresponds one state of equilibrium, to which the system evolves. Because of this uniqueness, it is possible to describe any state of equilibrium in terms of the constraints that bring it about. Indeed, equilibrium states are described in terms of constraints: the magnitude called "volume," for example, describes the space to which the system is confined; the magnitude called "pressure" is determined to be equal to the force exerted on the system; and the temperature of the system is determined by the external environment. However, since the thermodynamic magnitudes are described in terms of the constraints, which are by definition associated with equilibrium, standard thermodynamics cannot describe non-equilibrium states and fluctuations away from equilibrium. The reason for this use of constraints in describing thermodynamic states is clear from the history of thermodynamics: thermodynamics originated with the engineering problem of improving engines, and since the agent can affect the system only by way of imposing external constraints, thermodynamics is actually a study of this effect. However, despite the fact that the thermodynamic magnitudes are described in terms of the constraints, they characterize not only the agent acting on the system, but also the system itself. To see this, consider for example the ideal gas law, $PV = nRT$. This law says that for a given system (consisting of *n* moles of gas) if two external constraints are chosen (such as *P* and *V*), the third (*T*, in this case) is determined (*R* is a parameter): the external agent is *not* free to choose any *T*. This means that the system is such that only certain triads of numbers (*P*,*V*,*T*) are possible. If we understood *P*, *V*, and *T* as constraints rather than properties, the ideal gas law would look like a miracle.

By contrast, in mechanics the distinction between constraints and states is manifest. The states are not determined in terms of the constraints; the constraints determine ranges of possible states – the ranges that make up the accessible region. This difference will enable us to describe, in terms of mechanics, phenomena such as fluctuations, which are indescribable in standard thermodynamics.

3.4 Time evolution of mechanical states

As we said earlier, the limitations on a system determine its accessible region, which contains all the states compatible with these limitations. However, the limitations do not dictate which of the points in the accessible region is the system's actual state at each point of time. This is determined by the system's *dynamical equations of motion* and *initial state*, as follows.

The dynamical equations of motion of a system determine a set of *possible sequences* of states, which describe possible evolutions in time; the initial state of the system selects one such sequence as the *actual sequence*, and the two taken together determine which state on the sequence is the *actual state* at each point of time.

The equations of motion are applications of Newton's Second Law of Mechanics, $F = ma$, for the specific conditions of the system of interest. These conditions may include internal parameters of the system (such as the masses of the particles of which the system consists) and forces acting on the system (such as a gravitational field in which the system is placed, or its confinement to a container).[14] Given the internal parameters and internal as well as external forces, one can apply Newton's law and write down an equation of motion, which describes (or generates) a set of sequences of states that the system can have.

---

[14] Classical Newtonian dynamics includes only forces that pertain between elementary particles, namely gravitation and electricity, which are compatible with the velocity reversal invariance (see below). Forces such as friction, which seem to be incompatible with velocity reversal invariance, are supposed to be reducible to the mechanical forces; see Hutchison (1993) and Callender (1995).

Equations of motion are represented in the state space by *trajectories*. Trajectories are continuous sequences of points in the state space that correspond to the sequences of states generated by the equations of motion of a particular system (see Figure 3.1). The accessible region of a system is fully covered by trajectories.[15] Only one of the trajectories is *actual* in the sense that a segment of it represents the actual evolution of the system of interest. All the other trajectories (and the rest of the segments of the actual trajectory) are *counterfactual*: they are consistent with the external constraints and limitations on the system, but do not describe the actual evolution of the system of interest. The choice of the actual trajectory among the counterfactual ones is made by the initial state.

The initial state is some state in the accessible region, which the system happens to be in, at some designated point of time. Any state within the accessible region can become the initial state, and the determination of the actual initial state is part of the process of preparing the system of interest. Once the initial state is selected, feeding it into the equations of motion yields a trajectory, which is the actual trajectory. Moreover, if the initial state is indeed the state of the system of interest at some point of time, the trajectory segments before and after the initial state describe the past and future evolutions of the system. In that sense, the equation of motion of a system can take as input any state of the system at any given time, and it yields as output the state of that system at any other time. If the state of a system at a given time is represented by illuminating the point in the accessible region which represents that state, then one may think of the actual evolution of the system as a sequence of illuminations of consecutive points along the trajectory, at a constant rate.

Formally, the mechanical equations of motion are *partial differential equations*, and this entails two additional properties – endlessness and determinism – that are significant for underwriting thermodynamics by mechanics.

Endlessness: The trajectory of a mechanical system is infinite in both directions; it has neither a beginning nor an end. Thus any talk about initial conditions of the universe must be reduced to talk about some state of the system along its trajectory, or to talk about cutting the trajectory

---

[15] Not by a single trajectory, for topological reasons. This fact is important in the history of statistical mechanics since it undermines Boltzmann's original Ergodic Hypothesis.

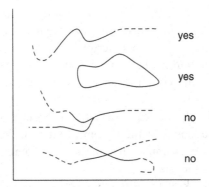

Figure 3.3 Trajectories consistent and inconsistent with determinism

by hand at some point, justifying this action by reference to other theories, such as the general theory of relativity which can accommodate notions such as the Big Bang.

Determinism[16]: The sequences of states generated by the equations of motion are such that if a state belongs to one sequence then it does not belong to any other sequence. In terms of state space, every point in the accessible region of the system sits on *exactly one* trajectory. Therefore trajectories never split, converge, or cross each other, as illustrated in Figure 3.3.[17]

This completes what one needs to know about mechanics in order to begin to see what it would mean to underwrite thermodynamics by mechanics. We now turn directly to take our first steps in this project: we will first introduce briefly the main conceptual tools of mechanics, namely dynamical blobs and macrostates (the latter notion is developed in greater detail in Chapter 5), and then use these tools to show why mechanics *cannot* underwrite the laws of thermodynamics, unless these laws are significantly weakened.

3.5 Thermodynamic magnitudes

In order to introduce the connection between mechanics and thermodynamics, consider the paradigmatic case of a gas in a container of

---

[16] There are various distinct notions of determinism in the literature, which we do not address here; see Earman (1986).
[17] If we restrict attention to subspaces of the total state space, the *projected* trajectories can cross, split, and appear almost in any way we want.

volume v, and recall the difference in the way in which each of the two theories describes this simple case. In thermodynamics a description of the external constraint is also a description of the state of the system subject to this constraint, when the system is in equilibrium: in our example the size of the container is v, and the volume of the gas in equilibrium is v as well.

By contrast, in mechanics the external constraint is compatible with *numerous* mechanical states of the system of interest, which is subject to this constraint; in our example, the container is compatible with many mechanical states of the gas.

Another way to look at the difference between the two theories is to notice that within mechanics two notions of states are involved, and only one of them corresponds to the thermodynamic state of a system. We call the first notion *microstate* and the second notion *macrostate*, and characterize these notions as follows.

Microstate: The microstate of a system at a given time is the positions and velocities of the particles making up that system at that time.

Macrostate: The set of microstates all of which appear to us as corresponding to the same thermodynamic magnitude is a macrostate of the system. The macrostate of a system is compatible with infinitely many microstates in the sense that some slight changes in the microstate of the system do not affect its overall appearance

Whereas the microstate of a system is represented by a point in the state space, a macrostate is represented by a region in the state space. This region is within the accessible region, and there are other subregions in the accessible region that are compatible with other macrostates. For example, the microstates in one region correspond to the macrostate in which a gas fills the entire volume of the container, and the microstates in another region would appear to us as the gas filling half of the container. In general, this characterization of macrostates corresponds to a partition of the state space of a thermodynamic system into mutually exclusive and exhaustive regions. Note that each thermodynamic quantity such as volume, pressure, and temperature yields a different partition of the state space into macrostates.

The partition of the state space into macrostates expresses not a property of the system (the gas, in our example), but *a property of the relation between the system and the observer* (us, in this example). The property that is expressed by this partition is the fact that the observer

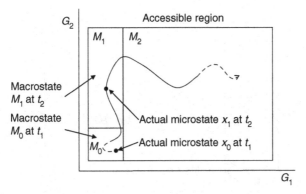

Figure 3.4 Macroscopic trajectory (superimposed on Figure 3.1)

cannot distinguish between the microstates of the system within any given macrostate. In Chapter 5 we will explain how this partition comes about, and show that this property, and the status of the observer in the theory in general, is not subjective or epistemic, but objective and physical. We come back to the notion of macrostates in more detail in Chapter 5.

Since the evolution of mechanical macrostates will be at the centre of our discussion of underwriting thermodynamics by mechanics, it is important to realize what exactly we mean by talking about the evolution of macrostates. In Figure 3.4 the system G is represented by two degrees of freedom, G_1 and G_2. The regions $[M_0]$, $[M_1]$, and $[M_2]$ are macrostates within the accessible region of G. (We denote regions of state space corresponding to macrostates in square brackets.) The partition of the state space into macrostate regions is fixed in time (the way in which it is determined is discussed in Chapter 5). As illustrated in Figure 3.4, the trajectory of G evolves in and out of the macrostate regions. To our eyes, this may seem like an evolution of the macrostate as such. For example, if macrostate $[M_0]$ in Figure 3.4 corresponds to the gas filling half of the container, macrostate $[M_1]$ to the gas filling three-quarters of the container, and macrostate $[M_2]$ to the gas filling the entire container, then the evolution depicted in the figure would appear to us as the gas *expanding* in the container. As long as the trajectory is within the region $[M_0]$, we will not notice any change in the state of the gas; we would notice a change only when the trajectory passes to region $[M_1]$ and then to $[M_2]$. The foregoing construction of macrostates in mechanics entails the following conclusion (which is of extreme importance in the underwriting of

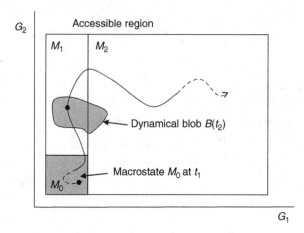

Figure 3.5 Dynamical blob: a dynamical evolution of a macrostate (super-imposed on Figure 3.4)

thermodynamics by mechanics): since the thermodynamic magnitudes correspond to mechanical macrostates, it is such a macroscopic mechanical evolution that should underwrite the laws of thermodynamics.

To better understand the idea of the time evolution of a macrostate, consider again our gas in a container (see Figure 1.1), and suppose that we have just removed a partition that confined the gas to half of the volume, so that our gas starts out in the non-equilibrium state in which it fills only half of the container. According to the thermodynamic Law of Approach to Equilibrium, the gas will evolve to fill the entire container. Let us describe this evolution in mechanical terms.

Our gas starts out at time t_1 in some microstate x_0 in the region of macrostate $[M_0]$. All the microstates in this region appear to us as the gas filling half of the container; see Figure 3.5. (Notice that this figure is superimposed on Figure 3.4, since both figures describe the same case of the gas expanding in a container.) At time t_1 *only* the microstate x_0 is actual, and all the other microstates in $[M_0]$ are counterfactual, in the sense that they are compatible with everything we know about the gas. Nevertheless, since only one microstate can be actualized at a given time, all the other counterfactual microstates do not happen to be the case. Since all we know is that the gas fills half of the container, we do not know *which* of the points in $[M_0]$ is the actual microstate.

The determinism of the classical equations of motion entails that each of the microstate-points in $[M_0]$ sits on a single trajectory (not necessarily

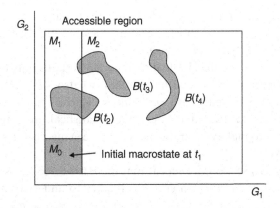

Figure 3.6 Liouville's theorem

on different trajectories). Let us now follow the evolution of all the points in $[M_0]$ – both the actual point x_0 and the counterfactual points – in accordance with the equations of motion, from time t_1 to some later time t_2. Each of the points that started out in $[M_0]$ at t_1 follows a well-defined trajectory segment that ends up at t_2 at some well-defined point, determined by the deterministic equation of motion. The end points of these trajectory segments make up the set of points $B(t_2)$, as illustrated in Figure 3.5. We call this set the *dynamical blob* of the system at time t_2.

The evolution of the dynamical blob is subject to an important theorem in mechanics proved by Liouville. According to Liouville's theorem the volume of the dynamical blob $B(t)$ at any time t must be equal to the volume of the initial macrostate $[M_0]$.[18] This can be seen intuitively by noting that each point in $B(t)$ is the end point of exactly *one* point in $[M_0]$ because of the determinism of the equations of motion.

Formally, in Liouville's theorem what we called here the *volumes* of $[M_0]$ and $B(t)$ are given by their *Lebesgue measures* (which is a generalization of the intuitive notion of size; we expand on the notion of measure below).

While Liouville's theorem determines the volume (by the Lebesgue measure) of the dynamical blob, it places no constraint on its *shape*. Figure 3.6 illustrates that, for example, the set of points that start out at t_1 in $[M_0]$ may evolve through $B(t_2)$ at t_2 to $B(t_3)$ at t_3 and then to $B(t_4)$ at t_4. The term *dynamical blob* is intended to express the fact that the shape

[18] See Landau and Lifshitz (1980) for a full proof, and Albert (2000, Ch. 3) for an informal sketch of the proof of Liouville's theorem.

of these regions may change radically, while the dynamics conserves their Lebesgue measure.[19] (For short, we sometimes use the term *blob* rather than the full name dynamical blob.)

It is important to see that Liouville's theorem pertains only to the time-dependent *dynamical blob*, and has nothing whatsoever to do with the partition of the state space into *macrostates*, which is time-independent. Some confusion in the literature about statistical mechanics arises through failure to make a clear distinction between macrostates and blobs in this context.

To sum up this section, we repeat that there are three kinds of regions in the state space, and their interplay is the key to underwriting thermodynamics by mechanics (see Figure 3.5). (i) The accessible region contains all the points compatible with the constraints and limitations on the system. It is stationary in time, for as long as the constraints are unchanged. (ii) The macrostates are determined by the fact that certain sets of microstates appear to have the same thermodynamic magnitudes; The partition into macrostates is stationary. (iii) The dynamical blob at a given time is the set of end points to which the points of the initial macrostate have arrived at that time. The blob is determined by the dynamics of the system: given that the system starts out in some macrostate $[M_0]$ at time t_1, the equations of motion determine its dynamical blob at any other time.

3.6 A mechanical no-go theorem

In the previous sections we introduced conceptual tools with which it is best to think about mechanics and presented them in terms of three kinds of state space regions: the accessible region, macrostates, and dynamical blobs. We now proceed directly to using these tools for the task of underwriting the laws of thermodynamics by mechanics. As we will see, this will immediately reveal an important no-go theorem, according to which it is impossible to underwrite the strict and absolute laws of thermodynamics by mechanics, and replacing them by some weaker probabilistic counterpart is theoretically necessary. This result is encouraging, since, as we said towards the end of Chapter 2, the need to account

---

[19] The notion of dynamical blob is close to the so-called *Poincaré sections* used to characterize periodic and semi-periodic systems.

for the phenomenon of fluctuations leads to a similar conclusion. In this section we wish to outline this no-go theorem in order to show why mechanics by itself is insufficient to underwrite thermodynamics. For this purpose we need to show how a mechanistic counterpart of the Law of Approach to Equilibrium and the Second Law would look, and to this end we shall focus on the notion of entropy (later on, in Chapters 5 and 7, we give a fuller mechanical account of thermodynamics).

We saw above that thermodynamic properties such as volume, pressure, and temperature are represented by macrostates (see Chapter 5). If the microstate of a system at some given point of time is within the region corresponding to that macrostate, then this system has the macroscopic properties corresponding to that macrostate; and if the system had been in any of the other microstates in that macrostate, it would have appeared to us as having the same thermodynamic properties. Now, in thermodynamics, entropy is also a way of characterizing a system's thermodynamic state. The property of entropy is slightly different from other properties in that it is relative: only entropy differences are defined. It is also different in that it is not independent, but is rather a function of the other properties; in an ideal gas, for example, the entropy difference is given by $\Delta S = c_V \ln(T_B/T_A) + R \ln(V_B/V_A)$, where c_V denotes the molecular heat capacity at constant volume, and R is a constant. However, the fact that entropy is a function of other magnitudes, which in turn are represented by macrostates, indicates that entropy itself (or entropy difference) should also be represented in terms of the macrostate of the system (or some difference between macrostates).

But *which* property of macrostates corresponds to entropy? In statistical mechanics (for reasons that we explain in Chapter 7) the mechanical property that corresponds to thermodynamic entropy is the *logarithm of the size* of the macrostate, as determined by the Lebesgue measure. This is Boltzmann's famous definition: $S = k \log W$, where S is the entropy, W is the Lebesgue measure, and k is a constant. (Hereafter we may occasionally talk about entropy as given by the size of a macrostate, but of course we always mean the logarithm of that size.) Intuitively, the idea here is that the larger the size of a macrostate, the larger the set of microstates in that macrostate, so the less we know about which is the actual microstate within that set, and therefore the less control we have over the actual microstate within that macrostate, and therefore, in turn, the less exploitable is the energy in order to produce work – which is what entropy is about in thermodynamics. Notice that this expression is about the entropy of a state, and not about entropy difference as in

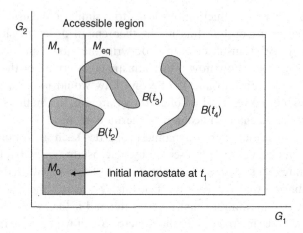

Figure 3.7 The mechanical counterpart of the Law of Approach to Equilibrium

thermodynamics. Moreover, this notion of entropy is defined for all macrostates, and not only for equilibrium macrostates as in thermodynamics. This is a conceptual generalization, involving a profound change that is brought about by replacing thermodynamics by mechanics (again see Chapter 7 for more details).

Given this definition of entropy, the Law of Approach to Equilibrium and the Second Law of Thermodynamics in their *strict non-probabilistic version* have the following meanings in mechanics.

Definition of equilibrium

Given the partition of the state space of a thermodynamic system into macrostates, one of the macrostates called $[M_{eq}]$ is identified with the thermodynamic state of equilibrium. $[M_{eq}]$ is uniquely determined by the constraints defining the accessible region. For example, in Figure 3.8 the set A of limitations and constraints determines the accessible region A, for which the equilibrium macrostate $[M_{eq}(A)]$ is uniquely determined; and the set B of limitations and constraints determines the accessible region B, for which the equilibrium macrostate $[M_{eq}(B)]$ is uniquely determined.

The Law of Approach to Equilibrium

Consider an isolated system starting out in *any* macrostate $[M_0]$ in its accessible region (see Figure 3.7). The dynamical blob $B(t)$ that initially coincides with $[M_0]$ invariably evolves over the accessible region in such a

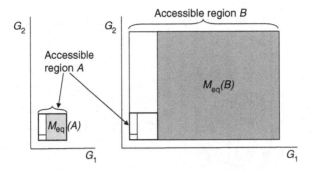

Figure 3.8 The mechanical counterpart of the Second Law

way that after some finite time interval it will be within the macrostate $[M_{eq}]$, where it will remain indefinitely.

The Second Law

Suppose that an isolated system is subject to the set A of constraints and limitations, which determines the accessible region A (see Figure 3.8). And suppose that the system starts out in the equilibrium state $[M_{eq}(A)]$ which is uniquely determined for A. Now suppose that the constraints on the system are changed, so that they make up the set B of constraints. This set determines the accessible region B, for which there is a uniquely defined equilibrium macrostate $[M_{eq}(B)]$. Then, it is invariably the case that the Lebesgue measure of $[M_{eq}(B)]$ is larger than or equal to the Lebesgue measure of $[M_{eq}(A)]$. If the mechanical counterpart of entropy is taken to be the Lebesgue measure of a macrostate, then the Second Law states that upon a change of constraints, the entropy is either conserved or increases, but never decreases. (We examine the connection between entropy and the Lebesgue measure in Chapter 7.)

Combining the Law of Approach to Equilibrium together with the Second Law, as stated above, means that a system subject to constraints A will be in the equilibrium state $[M_{eq}(A)]$, and as the constraints are changed to B the system will evolve to $[M_{eq}(B)]$, so that the evolution will be either entropy-conserving or entropy-increasing. Often this *combined* law is referred to as the Second Law. We emphasize the distinction between these two laws, since we believe that they require different kinds of explanations: the Second Law (as stated above) is about the structure of macrostates, and the Law of Approach to Equilibrium (as stated above) is about the evolution of the dynamical blobs.

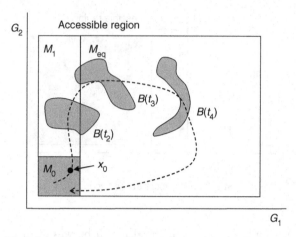

Figure 3.9 Poincaré's recurrence theorem

A strict derivation of the time-asymmetric laws of thermodynamics from the laws of mechanics is logically impossible, since the latter are time-symmetric (see Chapter 4). This point can be illustrated in several ways in terms of the mechanical concepts presented so far. Here are three such ways.[20]

(i) Consider Figure 3.7. According to the above description of the Law of Approach to Equilibrium, whether the system starts out in $[M_0]$ (as in the figure) or in $[M_1]$ or $[M_{eq}]$ (cases not depicted in this figure), it should end up in $[M_{eq}]$.[21] In other words, whether the dynamical blob $B(t)$ of the system starts out as coinciding with $[M_0]$, $[M_1]$, or $[M_{eq}]$, it should end up within $[M_{eq}]$ and remain there indefinitely. This means that the *entire* accessible region is mapped to $[M_{eq}]$, and this is a case in which a region evolves to a smaller region, in violation of Liouville's theorem.

(ii) Another way to see why this law as stated is incompatible with mechanics is via another theorem of mechanics proved by Henri Poincaré, called the *Poincaré recurrence theorem* (see Figure 3.9). According to this theorem the trajectory of a system starting from *any* initial microstate, call it x_0 (except for a set of points of Lebesgue

---

[20] Note that the following proofs are presented in a non-historical way. For the history of these ideas, see Sklar (1993) and Uffink (2007).

[21] Notice that in this figure the size as determined by the Lebesgue measure of M_{eq} is substantially larger than the other macrostates, in this case M_0 and M_1. This is indeed customary in statistical mechanics, but these differences in size need be demonstrated; we discuss this issue in Chapters 5 and 7.

measure zero), will arrive at some other time to *any* region around x_0 which is of positive Lebesgue measure. Since the region of the macrostate that contains x_0 as it starts its evolution ([M_0] in the figure) is of positive Lebesgue measure, the trajectory that starts out in x_0 will, at some point of time, leave the equilibrium state [M_{eq}] and return to [M_0], in violation of the Law of Approach to Equilibrium. Poincaré's recurrence theorem was brought up by Ernst Zermelo as an objection to Boltzmann's so-called *H*-theorem, in which Boltzmann attempted to derive the Law of Approach to Equilibrium from the mechanical equations of motion. The usual reply to the Poincaré recurrence objection is that the recurrence time, after which the system will return close to x_0, is extremely long, so that we are unlikely to experience its effect. This reply is in line with the weakening of the laws of thermodynamics by replacing them with probabilistic counterparts – an idea that we examine later.

(iii) The third argument is based on Josef Loschmidt's so-called *reversibility objection* to Boltzmann's *H*-theorem, and is often mentioned alongside Zermelo's use of the Poincaré recurrence theorem, described above. Consider any microstate x_0 that belongs to the macrostate [M_0] and that evolves in the course of time, after some time interval Δt, to the microstate x_1 in the equilibrium macrostate [M_{eq}], in accordance with the Law of Approach to Equilibrium. Loschmidt pointed out that for any such microstate x_1 in [M_{eq}], there is a different microstate, call it $-x_1$, which is also in [M_{eq}], and which is identical to x_1 in its position coordinates but is the exact velocity reversal of x_1. By *velocity reversal* we mean that the magnitude of the velocity of the system in $-x_1$ is the same as in x_1, but its spatial direction is exactly the opposite: if the velocity in x_1 travels leftwards or upwards, the velocity in $-x_1$ is rightwards or downwards. An important property of the mechanical equations of motion (on which we expand in Chapter 4) is that if the system starts out in $-x_1$, it evolves after Δt to the microstate $-x_0$, which is the velocity reversal of x_0 and is within [M_0].[22] The evolution from $-x_1$ in [M_{eq}] to $-x_0$ in [M_0] violates the Law of Approach to Equilibrium; and since to any evolution from x_0 in [M_0] to x_1 in [M_{eq}] – which obeys the Law of Approach to Equilibrium – there corresponds an evolution from $-x_1$ in [M_{eq}] to $-x_0$ in [M_0] that violates this law, it turns out that

---

[22] This means that the macrostate of a system is taken to be invariant under velocity reversal; we expand on this idea in Chapter 7.

evolutions that violate this law are as common – or as rare – as evolutions that satisfy it. Boltzmann's reply to this objection was given in terms of conditional probabilities; we will address this idea in Chapter 7.

These proofs show that thermodynamics cannot be underwritten by mechanics alone. Following Boltzmann, the prevalent approach in statistical mechanics addresses the above three objections by appealing to the notion of *probability*. It is the appeal to probability which opened the door to statistical mechanics. Roughly, the idea is that an evolution that satisfies the Law of Approach to Equilibrium and the Second Law of Thermodynamics is highly likely to such a degree that we ought never to expect to see their violation. We now turn to discuss in outline one important direction in the prevalent approach to statistical mechanics, which follows this strategy.

3.7 The ergodic approach

The ergodic approach in its modern version[23] is one of the important attempts at underwriting thermodynamics by mechanics in view of the above no-go theorems.[24] The main idea is to replace the mechanical version of the Law of Approach to Equilibrium by a probabilistic counterpart which would be *derivable* from the mechanical equations of motion. This derivation is carried out in a special case (which is taken to be characteristic of thermodynamic systems) of the so-called ergodic dynamics, in which, roughly, the relative time that a system spends in a region of the state space is proportional to the size (as measured by the Lebesgue measure) of that region. That is, the idea is to arrive at something like the following statement: An isolated system starting out in any microstate in any macrostate in its accessible region, evolves in such a way that at any time the system is observed, it is extremely highly likely to be found within the equilibrium macrostate [M_{eq}]. If something like this statement could be proved, it would entail that a system that starts out in any non-equilibrium macrostate would, with high probability, evolve

---

[23] Including the idea of ε-ergodicity, see Frigg and Werndl (2012). For a general background and history, see Arnold and Avez (1968), Sklar (1993), Uffink (2007) and Berkovitz, Frigg and Krontz (2006) and references therein.

[24] But see Earman and Redei (1996) for criticism regarding the significance ascribed to ergodicity in statistical mechanics.

to equilibrium, and a system that is already in equilibrium would, with high probability, stay there; and this seems like a mechanical underwriting of a probabilistic counterpart of the Law of Approach to Equilibrium.

Notice that in the formulation of the above statement we do not mention the time interval Δt of approach to equilibrium, which plays an important role in the above formulation of the Law of Approach to Equilibrium. Moreover, we do not mention the location of the dynamical blob $B(t)$ in the state space relative to $[M_{eq}]$. Both omissions are intended, and we will discuss them later.

This widely accepted approach is expressed by Hans Reichenbach as follows.

> There still remained the question whether the ergodic hypothesis must be regarded as an independent presupposition, or whether it is derivable from the canonical equations [of motion], as Liouville's theorem is. If the first alternative were true, we would have here a derivation ... in which a probability assumption, the ergodic hypothesis, is added to causal laws, namely Liouville's theorem, in order to achieve the derivation of the probability metric. If the second alternative were true, we would have here a derivation of a probability metric from causal laws alone. This problem has occupied mathematicians for a long time. It was finally solved through ingenious investigations by John von Neumann and George Birkhoff, who were able to prove that the second alternative is true ... With von Neumann and Birkhoff's theorem deterministic physics has reached its highest degree of perfection: the strict determinism of elementary processes is shown to lead to statistical laws for macroscopic occurrences, the probability metric of which is the reflection of the causal laws governing the path of the elementary particle.[25]

It is fair to say that Reichenbach is expressing here, quite accurately, the ergodic approach which is still quite prevalent in statistical mechanics. We will now show why this approach is mistaken, focusing on two main issues. The first is that, *contra* Reichenbach, it is not at all evident that the ergodic theorem is in fact a theorem about probability. The other reason is that ergodicity, to the extent that it entails anything about probability, does so in a way which is *empirically empty* since it entails no predictions whatsoever about any possible experience.

Let us now present the way in which the ergodic approach argues for its claims.

(I) *The first assumption* in the argument is that the size of $[M_{eq}]$ is by far greater than the size of the other macrostates, so that $[M_{eq}]$ takes up

---

[25] Reichenbach (1956, pp. 78–79).

almost all of the accessible region. Size here is determined by the Lebesgue measure. This assumption seems reasonable given certain paradigmatic examples (but it is by no means trivial, and we examine it in Chapter 8). The aim in the ergodic approach is now to prove that the relative frequency of a macrostate along the trajectory of a system is determined by the Lebesgue measure of that macrostate and therefore the system spends most of the time in equilibrium.

(II) To achieve this aim the ergodic approach implicitly makes a linkage between probability, measure and relative frequency as follows. There is a set of points in the phase space of an ergodic system all of which sit on trajectories such that the relative time a system spends in each macrostate is proportional to the Lebesgue measure of that macrostate (we later call this set BvN). It is proved that the Lebesgue measure of this set is 1 (this is the proof by Birkhoff and von Neumann Reichenbach appeals to). Note that the Lebesgue measure here has two different roles, and we now focus on its second appearance. It is assumed (implicitly) in the ergodic approach that Lebesgue measure is probability. And therefore it follows that with probability 1, the relative frequency in time of a macrostate along the trajectory of a system is equal to its Lebesgue measure. The crucial assumption in this argument is the leap from Lebesgue measure 1 to probability 1. As we shall see, Birkhoff and von Neumann's proof indeed establishes the linkage between Lebesgue measure and relative frequency (in the infinite time limit) but their theorem does not establish a connection between Lebesgue measure and probability (we expand on this important point in Chapter 8). One way to think about this distinction is the following. Suppose that we observe an ergodic system at a randomly chosen point of time t, at which the system is in some microstate x. Since t is chosen at random, x is also chosen at random out of the accessible region. Thus, metaphorically and intuitively, observing a system at a random point of time is tantamount to sampling at random a point out of the accessible region of the system. To say that a point is chosen at random means that all points are equally likely to be chosen, and therefore in a random sampling of a point out of the accessible region, the *probability* of picking out a point that belongs to any region R is given by the *size* of that region. (Here, again, size is determined by the Lebesgue measure; we discuss this choice below.)

(III) *The third assumption* is that events which have very small probability never occur, and events with very high probability occur with certainty; this idea is known as *moral certainty*.[26]

(IV) *The conclusion* from the former assumptions is as follows: since according to (I) the equilibrium macrostate takes up almost the entirety of the accessible region, according to (II) this macrostate has very high probability, and therefore according to (III) the system is in equilibrium with certainty at any time. And so if the system happens to be in some non-equilibrium state, it will, with certainty, approach equilibrium and remain there. This sounds like a mechanical probabilistic counterpart of the thermodynamic Law of Approach to Equilibrium.

(V) To this conclusion one may add the idea that the mechanical counterpart of the thermodynamic notion of entropy is the size (by the Lebesgue measure) of the macrostate of that system. (We discussed this idea earlier in this chapter, and will return to it in Chapter 7.) And so the Lebesgue measure of a macrostate has, in the ergodic approach, a double role: the larger the size of a macrostate, the higher is the *entropy* of that macrostate, and the higher is the *probability* that the system will be found in this macrostate upon a random sampling. Hence, the idea that the ergodic systems have a high tendency to increase their entropy sounds almost tautological. We do *not* take this to be a fourth step of the ergodic approach, for the following reason: we take ergodicity to be an argument for the Law of Approach to Equilibrium, and the notion of entropy plays no role in this law. It appears only in the Second Law of Thermodynamics, as defined above in this chapter. (For this reason, hereafter in this section we will avoid mentioning the notion of entropy.)

But none of the above assumptions in the argument is trivial, and in fact they are not general and often false. The main problem underlying the ergodic approach is assumption (II). As we will now argue (and expand in several places throughout the book), there is no reason to suppose that the Lebesgue measure of a macrostate has anything to do with its probability; and hence even if one accepts (I) and (III), one does not reach the conclusions (IV) and (V). To see why this is so, let us see how the ergodic approach tries to establish assumption (II).

---

[26] Bernoulli (1713); Kolmogorov, (1933, Ch.1, §2).

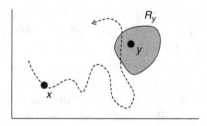

Figure 3.10 Ergodic dynamics

The intuitive idea behind the ergodic dynamics is that if the trajectory of a system evolves, as it were, randomly all over the accessible region, then it seems plausible to say that the system is equally likely to be found anywhere in the accessible region, and therefore it seems plausible that the probability of finding the system in a macrostate is proportional to its size. Indeed, in order to support a probabilistic conjecture such as the one in step (II) above, Boltzmann conjectured that thermodynamic systems evolve in such a way that their trajectories cover the whole accessible region; this conjecture came to be known as Boltzmann's *Ergodic Hypothesis*. But Boltzmann's hypothesis is false, since it contradicts a theorem in topology according to which a one-dimensional line cannot cover a multi-dimensional space.[27]

Boltzmann's hypothesis has been replaced by a theorem by Birkhoff and von Neumann, called the *quasi-ergodic* theorem (or metric-indecomposability theorem). The theorem proves a property of the trajectories of systems whose dynamics is such that they have no non-trivial constants of motion; we will call such systems and such dynamics *ergodic systems* and *ergodic dynamics*; and their property is as follows. The accessible region of an ergodic system is divided into two disjoint sets of points. One set (call it *BvN* after Birkhoff and von Neumann) is such that a trajectory that starts out in any of its points passes arbitrarily closely to any other point in the accessible region. In other words (see Figure 3.10): take *any* point x in the *BvN* set and *any* point y in the accessible region (y does not have to belong to the *BvN* set); and surround the point y by some region R_y which can be as small as you like (where its size is determined by the Lebesgue measure); then, the trajectory starting out at x will pass through R_y at *some time*. The second set, the complementary of the *BvN* set (call it the *non-BvN* set) is such that for any point z in it, there is a region R (z) in the accessible region, with positive (non-zero) size (where again this

---

[27] This was shown by Rosenthal and Pluncherel in 1913; see Sklar (1993, p. 77).

size is determined by the Lebesgue measure) such that the trajectory starting out in z will *never* pass through $R(z)$.[28] And the size of the *BvN* set (as determined by the Lebesgue measure) is 1, and the size of the *non-BvN* set (as determined by the Lebesgue measure) is 0.[29]

This result has the following consequence, which gives the theorem its significance in the attempts to construct probabilistic mechanical counterparts of the thermodynamic Law of Approach to Equilibrium. Take a trajectory that starts out in a point that belongs to the *BvN* set (and this entails that all the points on this trajectory belong to the *BvN* set); for short, let us call this a *BvN* trajectory. Now, consider any region R_y which has positive (non-zero) size (as determined by the Lebesgue measure). Since we know from the theorem that the *BvN* trajectory will with certainty pass through R_y, we can follow this trajectory, along its infinite evolution, and mark all its segments that are within R_y, and add them up, and determine the relative size (by Lebesgue measure) of this sum of segments. Now, since the trajectories in the state space represent the mechanical equations of motion, the system evolves *in time* in constant rate along this trajectory, so that equal *lengths* (by Lebesgue measure) of trajectory segments are traversed during equal *time* intervals. And for this reason we can say that the sum of *lengths* of trajectory segments that pass through R_y expresses the *time* that the system spends in region R_y. In particular, if the region R_y corresponds to some macrostate of the system, then the sum of the *lengths* of the trajectory segments that pass through this macrostate expresses the relative *time* that the system has the thermodynamic property corresponding to this macrostate. Given this understanding of the connection between time intervals and trajectory segments, the Birkhoff–von Neumann theorem entails that *the relative time that a system spends in a macrostate is given by the Lebesgue measure of that macrostate.* This property holds only for ergodic trajectories, that is for a set of measure 1 of points in the state space. *Trajectories that start out in the non-BvN set of points do not exhibit this property.* This statement is sometimes called the ergodic theorem, and the property itself, because of its importance, is sometimes referred to as *ergodicity*.

---

[28] One can see immediately that the BvN theorem entails the Poincaré recurrence theorem. However, the converse is not true: that is, Poincaré recurrence holds even when the dynamics is not ergodic, and even if the trajectory of the system is confined to a proper subregion of the accessible region, as for example in the case of the KAM systems discussed below.

[29] There is a set of measures that are *absolutely continuous* with the Lebesgue measure, and these measures agree with the Lebesgue measure on the sets of points that have measures 0 and 1; hereafter, when we talk about the Lebesgue measure, we refer also to these measures, and when we talk about different measures we refer to those that are *not* absolutely continuous with the Lebesgue measure.

According to the theorem by Birkhoff and von Neumann, the *BvN* set of points is of Lebesgue measure 1, and the non-*BvN* set is of Lebesgue measure 0. Intuitively, this means that *most* of the points in the accessible region of an ergodic system are in the *BvN* set while a *minority* are in the non-*BvN* set. If we want this fact to entail that the *actual* microstate of a *given* ergodic system belongs to the *BvN* set, we must add the two following assumptions: that Lebesgue measure 0 entails probability 0 and Lebesgue measure 1 entails probability 1, which is the heart of assumption (II) above; and that for all practical purposes, events which have probability 0 are impossible, and those which have probability 1 are certain, which is assumption (III) above. These assumptions are necessary (though not sufficient) in order to render the Birkhoff and von Neumann theorem significant for underwriting thermodynamics by mechanics. *Given* these two assumptions, the theorem entails that if we observe an ergodic system at arbitrary times, we can be certain that its microstates at the times of observation will belong to the *BvN* set. From (I) we can conclude that the system is at equilibrium. And this seems to be a very reasonable probabilistic underwriting of the thermodynamic Law of Approach to Equilibrium.

However, although this result seems very elegant and hence attractive, it turns out that the significance of the theorem of Birkhoff and von Neumann in underwriting the laws of thermodynamics by mechanics is rather limited.

To start with, notice that the minority set of points, of the non-*BvN* set, need not be a small finite set of points, and is in general an infinite set. By comparison, the Lebesgue measure of the set of rational numbers among the real numbers is 0 (they are denumerable, $\aleph_0$) and the Lebesgue measure of the complementary set, of the irrational numbers, is 1 (they are non-denumerable, $\aleph_1$). Moreover, the rational numbers are dense among the real numbers; that is, between any two real numbers there are infinitely many rational numbers. By analogy, the set of non-*BvN* points can be infinite, and its points can be dense among the *BvN* points in the state space of an ergodic system. This means that a set of measure 0 is not necessarily small in any intuitive sense.

As we said earlier, in the context of discussing Liouville's theorem, the intuitive notion of size is generalized by the notion of a *measure*: the definition of a measure includes several characteristics of the intuitive notion of size; but it turns out that there is a variety of measures, all of which satisfy these conditions and can therefore serve in the role of size. In particular, a continuous set, such as the set of points in the state space, can be measured in a variety of ways.[30] For example, the distance between the two points x and y can be measured by $|x - y|$ or by $|x^2 - y^2|$.[31] Both functions satisfy the mathematical definition of measure, but they yield a different size for the distance between x and y. The *standard* way of measuring distance, area, and volume is by using the so-called Lebesgue measure, of which $|x - y|$ is a special case.

In the Birkhoff–von Neumann theorem, the Lebesgue measure features twice: in determining the size of the sets R_y and of $R(z)$, and in determining the size of the *BvN* and non-*BvN* sets. The significance often ascribed to the theorem hinges on the fact that these sizes are as determined by the Lebesgue measure. If size were to be determined according to a measure which is very different from the Lebesgue measure, then the theorem would lose its significance. The question is, then, why use the Lebesgue measure, rather than any other measure?

Needless to say, the fact that the theorem is proved for the Lebesgue measure cannot be a criterion for preferring this measure, since the very idea that the theorem is significant hinges on the idea that one has independent reasons for preferring the Lebesgue measure; if one has no such *independent* reasons, the mere fact that the theorem proves elegant properties about this measure has no significance at all for underwriting thermodynamics by mechanics.

The circularity of the usual arguments for preferring the Lebesgue measure in this context can be seen clearly from the following line of reasoning. As we said earlier in this chapter, Liouville's theorem says that the Lebesgue measure is conserved under the mechanical evolution: in the above terminology, the Lebesgue measure of the dynamical blob is conserved, although the shape of the blob may radically change during its evolution. This fact is often brought up as a reason for taking the Lebesgue measure to be natural for measuring sizes in the mechanical state space. However, sizes of state space regions have different theoretical roles. The fact that the Lebesgue measure has a role in one context does not mean that it must be preferred in different contexts. In particular, Liouville's theorem is not relevant for choosing the Lebesgue measure in order to ascribe *probability* to macrostates (we expand on this point in Chapters 6 and 8, where we introduce a different notion of probability). True, if probability were to be given by the Lebesgue

---

[30] This idea is illustrated by famous paradoxes that show that measure on a continuous space is not unique. See for example van Fraassen (1989, Ch. 12).

[31] See Albert (2000, Ch. 3).

measure, then by Liouville's theorem the probability ascribed to a state space region would be conserved under the dynamics, as the blob starting out in this region evolves; but while such a property is perhaps elegant and convenient, it is not necessary and compelling in the context of questions about probability, and in particular about the relative frequencies of observed events. In other words, there is no reason why nature should be like that, or why mechanics entails that nature is like that. (We come back to the problem of preferring a measure in Chapters 7 and 8.)

Moreover, while Liouville's theorem says that the Lebesgue measure of a dynamical blob is conserved under the mechanical evolution over time, it does not imply that this measure is the only one with this property. However, such a unique property is obtained for ergodic systems: if a system is ergodic, then the only measure conserved under the dynamics is the Lebesgue measure (and measures absolutely continuous with it). However, the fact that the ergodic theorem entails such an elegant property of the Lebesgue measure does not constitute an independent reason for preferring the Lebesgue measure over other measures, and hence does not lend additional significance for appealing to this measure in the attempt to underwrite thermodynamics by mechanics.

There are two well-known additional points that undermine the ergodic approach.[32]

One is that to the extent that ergodic dynamics implies anything about the probability of macrostates, this entailment holds only as time goes to infinity. Even if we know that a system is ergodic, and even if we know that the trajectory of the system is a *BvN* trajectory, nothing at all in the theorem suggests that the probabilities of the different macrostates during any finite timescale has anything to do with the Lebesgue measure of the macrostates. Indeed, we see no argument at all to this effect, nor even very general plausibility argument. In fact, we suspect that there are strong arguments to the contrary, namely that the probabilities of macrostates are *not* proportional to their Lebesgue measure.

Here is one such very simple argument. Consider a system that starts out in some non-equilibrium macrostate, say $[M_0]$ in Figure 3.7. Then, if it were the case that the probability of $[M_1]$ and $[M_{eq}]$ in this figure had been proportional to their Lebesgue measure, the system would have evolved, with high probability, directly to $[M_{eq}]$, and would have never passed through $[M_1]$, contrary to our everyday experience concerning the gradual approach to equilibrium. Any claim to the effect that probabilities ought

---

[32] For example, see Earman and Redei (1996), Sklar (1993).

to refer to long-term evolutions, rather than short-term ones, ought to be rejected on the ground that long term has still to be in orders of magnitude that are relevant to empirical observations, if our theory is to be physical at all. To put it bluntly, the ergodic approach, to the extent that it purports to account for the Law of Approach to Equilibrium (or the Second Law, for that matter), is not physical.

The second point often mentioned in the literature, but not taken as a reason to abandon the ergodic approach, is that thermodynamic systems, such as Carnot's heat engine, are in general not ergodic. Often one encounters the opinion that they are; but no proof to this effect is available. On the contrary, there are reasonable grounds to suspect that thermodynamic systems are so-called KAM systems (characterized by Kolmogorov, Arnold, and Moser), that is, systems that have non-trivial constants of motion which keep their trajectories in regions of the state space that have positive Lebesgue measures.[33] Given the above arguments, especially the argument concerning the desideratum that the mechanical counterpart will account for the fact that the approach to equilibrium is achieved in a finite time and is empirically observable, we see no reason to adhere to the hope of proving that thermodynamic systems are ergodic. Such a proof, even if it were given, would contribute nothing at all to underwriting thermodynamics by mechanics.

3.8 Conclusion

In this chapter we went through some features of classical mechanical systems. We introduced the notions of accessible region, macrostates, and dynamical blobs, and using them we presented the mechanical counterparts of the Law of Approach to Equilibrium and the Second Law of Thermodynamics. After proving a no-go theorem, according to which these counterparts are contradicted by mechanics, we turned to examine whether probabilistic counterparts can be constructed. Here we presented one prevalent approach to the problem, namely the ergodic approach, and explained why we think this approach ought to be rejected as irrelevant to the task in hand. Our arguments here will serve as a basis for presenting an alternative approach to the problem. In Chapter 2 we emphasized that thermodynamics is often taken to provide an arrow of time. In the next chapter we will present the feature of mechanics that seems to clash directly with this arrow, namely, the time symmetry of classical mechanics.

---

[33] See Walker and Ford (1969), Sklar (1993).

4

Time

4.1 Introduction: why mechanics cannot underwrite thermodynamics

In Chapter 2 we introduced the thermodynamic arrow of time: given an unordered set of thermodynamic states that are known to have been taken from a single process that has occurred in time, it is easy to arrange these states in a sequence according to their temporal order (that is, the order in which they occur in time), since the Law of Approach to Equilibrium points to equilibrium states as being later than non-equilibrium states; moreover, the Second Law of Thermodynamics points to high-entropy equilibrium states as being later than lower-entropy equilibrium states. For instance, it would be easy to arrange the four parts of Figure 1.1 according to the right temporal sequence, from the early state *a* to the later state *d*. In attempting to underwrite the laws of thermodynamics by mechanics, one naturally expects to be able to reproduce this temporal ordering using mechanical terms. However, this task proves difficult, since although a *d*-to-*a* evolution violates the laws of thermodynamics, it is perfectly consistent with the laws of mechanics: both an *a*-to-*d* evolution and a *d*-to-*a* evolution satisfy the mechanical equations of motion. Mechanics does not select a unique temporal order of thermodynamic states. This lack of uniqueness is a consequence of several symmetries in mechanics, all of which are associated with time. Owing to this lack of uniqueness, it is impossible to recover the thermodynamic ordering of events in time from the mechanical ordering of events in time.

This is the general point we raised before (in Chapter 3): any attempted proof of the time-directed laws of thermodynamics from the laws of mechanics alone is *logically invalid*. At the same time, it seems that there is no logical contradiction between thermodynamics and mechanics: there are cases where both are true. And so the question is this: what assumptions need to be added to mechanics in order to pick out these cases? That is, what are the circumstances in which the mechanical principles and laws of motion give rise to the thermodynamic regularities? This is the central challenge of statistical mechanics.

In order to understand this challenge, it is important to understand what symmetries of mechanics are associated with time ordering. Some of these are presented in this chapter: we will focus here on the symmetries of classical *kinematics*, that is, of the theory of motion without forces. A full account of the symmetries of mechanics would have to address the time symmetries of the mechanical forces as well; we do not undertake this here.[1] It seems to us that an understanding of the time symmetries in kinematics is enough to grasp the problems involved in underwriting thermodynamics by mechanics, and this is all we need here.

4.2 Classical kinematics

Classical mechanics describes the evolution of the states of mechanical systems. Classical kinematics describes this evolution in cases where no forces are in action and, therefore, according to Newton's laws of motion, no acceleration takes place and the velocity is constant. The kinematic equation of motion of a point-like particle which moves along the x direction between the instants t_0 and t_1 is as follows:

$$x(t_1) = x(t_0) + \int_{t_0}^{t_1} \frac{dx}{dt} \, dt, \tag{1}$$

---

[1] The question of time symmetry in mechanics is also discussed in the context of classical dynamics, where forces are in action and the law of motion is Newton's Second Law $F = ma$ where $F \neq 0$. Here the question concerns the symmetry of the forces themselves and how it is related to the time symmetry. There are two main issues. The first concerns the magnetic force. For example, in order to return to the initial position, velocity reversal (in the sense discussed below in this chapter) is not enough. One also needs a reversal of the orientation of the magnetic force; see Albert (2000, Ch. 1), and reactions by Earman (2002), Malament (2004), Arntzenius (2004), and Arntzenius and Greaves (2009). The second issue is that of friction forces, and whether they are part of mechanics. See Hutchison (1993) and Callender (1995). Our discussion focuses on the kinematic case only, which suffices for our purposes.

where in the kinematic case the velocity $v = \mathrm{d}x/\mathrm{d}t$ is constant, which means that for all instants τ between t_0 and t_1, the instantaneous velocity at τ is

$$\left.\frac{\mathrm{d}x}{\mathrm{d}t}\right|_{\tau} = \lim_{t \to \tau} \frac{x(t) - x(\tau)}{t - \tau} = v_0, \tag{2}$$

so that (1) can also be written as

$$x(t_1) = x(t_0) + v_0(t_1 - t_0). \tag{3}$$

Equation (2) means that since in classical mechanics the instances of time are taken to form a continuum, instantaneous velocity (such as the velocity at the instant τ) is the limit of the difference between the positions $x(\tau)$ and $x(t)$ at the instances τ and t, as the difference between these instances approaches zero, and where this limit is the *same* on both sides of τ.[2] Equation (3) emphasizes the fact that Equation (1) of classical kinematics predicts the instantaneous position x of the particle at the instant t_1 if we are given the instantaneous mechanical state of the particle at the instant t_0, namely, if we are given the position $x(t_0) = x_0$ and the velocity $v(t_0) = v_0$ as given in (2).

4.3 The direction of time and the direction of velocity in time

Consider the film described in Chapter 1, in which scenes taken from a billiard game are shown. When we watch the film we experience the appearance of motion brought about by rapid succession of still frames. Each frame describes only the instantaneous positions of the billiard balls, and we experience as motion the change of these positions as the frames are shown one after the other. The sequence of events shown in this film is governed by the laws of mechanics, and the film itself provides a convenient metaphor for the description of mechanical evolutions in general. For the mechanical picture of the world may be described in terms of *time-slices*, analogous to frames of the film: each time-slice contains the positions of the particles at that instant, as do the film frames. In addition, the time-slices contain the instantaneous velocities of these particles, namely the limits as described in (2) above.[3]

---

[2] Note that the convergence from both sides of τ is a fact about the series of positions; it does not refer to the direction of time.

[3] Velocity is the rate of change of position with time, and so in order to define velocity one needs to take into consideration at least two instances, between which position changes; consequently it may appear that instantaneous velocity is an ill-defined concept. This

However, the value of v_0 is not determined only by the expression (2): this expression determines the magnitude of the velocity (known as the *speed*), but the sign of the velocity (that is, the *direction* of the velocity) is not given only by (2). In order to read off from (2) a unique description of the velocity of a particle at τ, we need to add to (2) information concerning the counterpart in physical space of the expression $x(t) - x(\tau)$, as well as information concerning the counterpart in physical time of the expression $t - \tau$. Normally this is assumed to be an easy task: the difference between two positions can be fixed by some convenient convention, such as $B > A$ if B is *to the right of* A; and the difference between two instances can be fixed such that, for example, $t_1 > t_0$ when t_1 is *later than* t_0. However, these conventions are not enough: in order to determine the signs of these differences uniquely, we need to specify a direction in space and a direction of time.

The sign of the spatial difference $x(t) - x(\tau)$ is determined by the meaning of the expression "to the right of"; and in order to specify this relation uniquely we need to add reference points in space.[4] Let us assume that we have such reference points, and so for every two points $x(t)$ and $x(\tau)$ along the x direction the difference $x(t) - x(\tau)$ is uniquely defined.

By contrast, it often appears that the sign of $t - \tau$ is determined by the expression "later than," which seems well defined. We always experience time as flowing in a direction that we call "from past to future," and this direction seems to determine the sign of $t - \tau$ uniquely. For this reason, there seems to be no need to mention explicitly the direction of time in any of the expressions above, and it seems that once the sign of $x(t) - x(\tau)$ is fixed by some convention, the velocity v_0 is well defined by (2). However,

intuition is famously expressed in Zeno's arrow paradox, according to which "the flying arrow is at rest, which result follows from the assumption that time is composed of moments ... If everything when it occupies an equal space is at rest, and if that which is in locomotion is always in a now, the flying arrow is therefore motionless" (Aristotle, *Physics*, 239b.30). Attempts to solve Zeno's paradox have led to various suggestions concerning the notion of instantaneous velocity, each of which involves difficulties in the statement of the laws of mechanics. (See description of the main types of approaches in Arntzenius 2000.) We shall not address Zeno's paradox and the attempts to cope with it in this book, and our use of the term *instantaneous velocity* will not involve committing ourselves to any of its currently prevalent interpretations. In this chapter we would like to point out a problem concerning instantaneous velocity, which is different from and additional to the set of problems that followed the attempts to deal with Zeno's paradox: the problem we wish to address concerns the connection between instantaneous velocity and the direction of time.

[4] The convention regarding space differences seems innocent and seems to carry no ontological baggage, even if one adopts an approach in which there is a preferred reference frame, such as Newton's approach expressed by Clarke: the preferred frame does not carry with it a preferred system of coordinates and of directions.

nothing in classical mechanics justifies this intuition. As far as classical mechanics is concerned, a universe in which time flows, as it were, from our future to our past, is conceivable. In Section 4.6 below we examine the meaning of a reversed direction of time; for now, it suffices to consider this idea in the most abstract way. In order for the velocity in (2) to be uniquely defined, one has to specify the direction *of* time. Otherwise the spatial differences in the numerator of (2) do not yield a well-defined velocity *in* time. And since instantaneous velocity is part of the instantaneous mechanical state of a particle, it turns out that *the direction of time is part of the instantaneous mechanical state of a particle*. We next address the way in which the direction of time can be described in mechanics.

4.4 The description of mechanical states

To see how the direction of time can be described in mechanics, we return to the billiard ball films described in Chapter 1. Each frame in the films describes the instantaneous position of each of the balls. But specifying these positions is not enough in order to single out the forward film or the backward one. The plots of the two films differ in the directions of the velocities of the balls. And so, in order to have a unique description of a film we need to add the order in which the frames are shown,[5] and this will determine the velocity of the particles. The specification of this order has two aspects: a relation we call *betweenness* and *direction*. We now turn to discuss them.

The first stage in specifying the order of frames is to fix the relation of *betweenness*, so that for each three frames we will be able to determine the order (f_i, f_j, f_k), in which the frame f_j is between frame f_i and frame f_k. As long as we are dealing with a stack of frames, which describe a discrete sequence of moments, the constraints on the order of stacks are relatively minor: there are many ways of ordering a stack of frames.[6] However, if we proceed from the stack of film frames to a sequence of the mechanical states of the billiard balls themselves (that is, the evolution that was filmed), further constraints must be added, due to the fact that the mechanical sequences of states is continuous, and to the fact that the balls have instantaneous velocities at each instant. The order of betweenness of the frames in a stack that describes a mechanical evolution is not

---

[5] And also the rate in time of their appearance – but we do not need to worry about this here.

[6] The constraints here express only the speed limit imposed by special relativity.

arbitrary: if we wish to have instantaneous velocities as defined in (2) at all times, the sequence of positions has to converge to the same limit on both sides of each frame.[7]

Given a suitable order of betweenness, we still are left with a directionless stack: the triad (f_i, f_j, f_k) can be either such that i is earlier than j which is earlier than k, or such that i is later than j which is later than k. Moreover: the betweenness relation can be read off from a given stack of frames; no additional external information is needed. But the relation of "*before*" and "*after*," which expresses a direction, cannot be read off from a stack of frames. If all we are given are such basic frames, then there is nothing from which we can read off a direction of the billiard film; the film is directionless. Notice that we do not argue that one direction is preferred over the other: there are two possible directions, and each showing of the film expresses one of them. But watching the stack of frames as such is not going to tell us the direction in which the film was filmed, and – more important – it is not going to tell us whether a particle shown in part of the stack was moving to the right or to the left when it was filmed. In order to have all this we have to add to the stack of frames an *external* arrow of time. This arrow cannot be read off from the stack of frames; it has to be provided in addition to them.

The billiard film frames are analogous to position degrees of freedom of particles. But a complete description of the mechanical state of particles contains instantaneous velocities as well. And a unique determination of instantaneous velocities assumes a direction of time, which cannot be read off from information concerning the positions of the particles. So it turns out that the very description of the instantaneous mechanical state of particles already assumes a direction of time, which is provided *in addition* to the positions of the particles.[8] To make this idea explicit, we add an arrow of time to each basic time-slice. (This idea can be implemented mathematically in various ways such as introducing a temporally orientable manifold.) Moreover, if we wish the described universe to reflect our experience, we need all the time arrows to point in the *same* direction; see Figure 4.1.[9]

---

[7] This means that the frames are not independent of each other, and so the picture described here is not Humean, not in the spirit of Russell's *at–at* approach. This is why we believe that the notion of an instantaneous state in mechanics implies not only a direction of time but also a passage of time.

[8] This idea is compatible with both substantivalism and with relationism about time, and here we do not take a stand in the debate between these two approaches.

[9] An alternative would be an arrow of time that stands outside the stack of frames, and somehow interacts with them. This alternative assumes from the outset that all the arrows must point in the same direction. This matches our actual experience but we do not wish to argue that different directions in different frames are inconceivable.

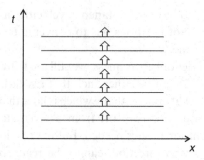

Figure 4.1 Time-slices

Such an arrow is the minimal thing we must add to each of the basic
frames, if we are to determine the order of the stack of frames and to
obtain a unique value of the velocities at that frame. If an arrow of time is
not added to each basic frame as an inherent part of it, there is nothing
that can encode such an arrow; nothing that can add a direction to the
relation of betweenness in order to construct a relation of "before" and
"after" and yield well-defined velocities.[10]

And so, it is here that the analogy between the mechanical picture of
the world and the films must end: the picture of the world according to
classical mechanics consists of a continuum of frames, each of which
contains not only positions but also an inherent arrow of time, so as to
add velocities to the elementary mechanical states of particles.[11] This
is the continuum of *time-directed time-slices*. We shall take *directed
time-slices* to be basic in our construction of mechanics.

The velocity of a particle at time τ is, then, determined *both* by the limit
$v_0 = \lim_{t \to \tau} \frac{x(t) - x(\tau)}{t - \tau}$ given in (2) *and* by the arrow of time built into the
time-slices. This arrow determines the sign of the denominator $t - \tau$ for
every t and τ, and is expressed in dt in the derivative dx/dt in (1) and (2).

In sum, in order to provide a full description of a sequence of classical
mechanical instantaneous states of a particle, one has to make the
following three independent decisions. First, one needs to fix the positions
of the particle within each time-slice, and choose a convention concerning

---

[10] It also seems to us that without such a built-in arrow, the physical transition (which is
something we experience) from one time-slice to the next – rather than to the previous
one – cannot be accounted for; but we shall not expand on this point here. We tend to
agree with Maudlin (2005, Ch. 4) on this point.

[11] Albert (2000, Ch. 1) seems to presuppose that position states are somehow embedded in
time, without explicitly imposing the direction of time into the basic frames. Without a
built-in arrow of time one cannot explain how the experienced time order of the frames
comes about, and how velocities assume their directions.

the way in which numbers denote these positions. Second, one needs to fix the order of the time-slices, so that they will satisfy a relation of betweenness. This ordering is subject to constraints if one wishes to end up with a universe in which instantaneous velocities exist. Third, one needs to fix a direction of time in each time-slice, in order to define the direction of velocity of the particle. To end up with a universe of the kind we experience, this direction has to be the same for all the time-slices. These three decisions are conceptually mutually independent. In practice, however, we make them on the basis of our experience, that is, the way we see particles in motion.

In the way we presented things above, the time-directedness of certain processes can be traced back to the fact that in our universe the arrows of time, which form part of instantaneous mechanical microstates, happen – as a matter of *contingent* fact – to point in a certain direction. Although the particular direction of time is fixed contingently, it is nonetheless an objective fact about our universe. An opposite direction is conceivable; but this possibility did not materialize in our actual universe.[12] This is comparable to the role and status of boundary conditions as acknowledged in classical mechanics. The equations of motion of classical mechanics (and other theories of physics) are differential equations, and such equations yield different predictions or retrodictions for different boundary conditions. For instance, given that at some instant the particles that make up a system have certain positions and velocities, the equations of motion predict and retrodict their positions and velocities at all other times. The fact that our actual universe – to the extent that it can be described by theories based on differential equations – was (or will be) at some instant in a certain instantaneous microscopic boundary condition, dictates everything else in the universe. And if this contingent fact is acceptable, then we think it is equally acceptable that the boundary condition that dictates everything else dictates also the direction of time.

In the kinematic Equation (1) the time parameter t appears in several roles, and we now turn to analyze these roles and consequently to distinguish between three different concepts: velocity reversal, retrodiction, and time reversal.

---

[12] Our view of an objective direction of time may be contrasted with what Huw Price (1996, Ch. 1) calls "a view from nowhen" according to which the temporal aspects of reality and in particular the direction of *time* are an appearance or an artifact of the way we describe certain evolutions in time.

4.5 Velocity reversal

Let us return to the example of the film of the billiard balls mentioned above. The film consists of a series of frames, each depicting the position of the balls at a specific instant. When we watch the film, we watch a series of frames in one of two orders: either in the forward order in which the frames were taken, or in reverse. But no matter in which order the film was taken or is shown to us, the direction of time in which we *watch* the film, the direction in which we experience the frames, is the same: it is the direction from the past to the future, which is the direction of all of our experience. The direction of time in which we experience the film is the same as the direction of all the other processes that we experience as we watch the film, such as the movement of the hands of clocks or some internal mental experiences, and of course it is the same as the direction of time in the actual billiard game that was filmed.[13]

The difference between the film shown in the forward order and in the reversed order is, then, not a difference in the direction of time: the difference is in the direction of the velocity of the billiard balls shown in the films. When we watch a film, we *interpret* the events that are recorded in the sequence of frames as events that occurred in the order in which they are shown; and so frames that are shown later are interpreted as events that occurred later in time. The result is that if a particle moved, say, from position A to position B when the film was taken, it will appear as moving from A to B in the forward-ordered film, and will appear to move from B to A in the reversed film. And this means that each run of the film is the *velocity reversal* of the other.

In terms of the kinematic equations of motion, velocity reversal is expressed as follows. Suppose that a particle moves along the x direction with constant velocity v from position $A = x(t_i)$ at the initial time t_i to position $B = x(t_f)$ at the final time t_f, and suppose that the forward time direction (which forms part of each time-slice or film frame) is from t_i to t_f. The particle then satisfies the following equation of motion, which is a special case of Equations (1)–(3) (see Figure 4.2):

$$B = A + \int_{t_i}^{t_f} \frac{dx}{dt}\, dt. \tag{I}$$

---

[13] We do not discuss here the question of how we sense the direction of time.

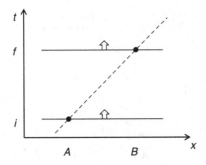

Figure 4.2 (I) A particle moving from A to B

In Equation (I) two conventions are involved: one concerning space and one concerning time. By the spatial convention, we set $B > A$, and this determines the value of any infinitesimal spatial distance dx in the derivative in (I). By the temporal convention, we set the forward direction of time, from past to future, to be denoted by increasing real numbers. Given that t_i is earlier in time than t_f, this convention entails that $t_i < t_f$. (Other conventions are possible, of course, and have their advantages as well as shortcomings.)

Now consider a film which shows the above evolution. If it is shown forward, then the frame *taken* at time t_i is *shown* at time T_i and the frame taken at time t_f is shown at time T_f, and if it is shown in reverse, then the frame taken at t_f is shown at T_i and the frame taken at t_i is shown at T_f. These films are described in Figures 4.3 and 4.4.

The sequence of events that we experience as we watch the forward film is described by an equation quite similar to (I), where the only difference is a time translation: from the time interval $[t_i,t_f]$ in which the film was taken to the time interval $[T_i,T_f]$ in which it is shown and the events in it are experienced:

$$B = A + \int_{T_i}^{T_f} \frac{dx}{dt} dt.$$

Let us now look for a mechanical description of the reverse-order film. As we have already said, the reverse film of the billiard balls seems to us as natural as the forward film, and indeed the evolution it depicts is possible by the theory of classical mechanics. The reversed sequence we seek (see Figure 4.4) is defined as the one in which the film begins at time T_i with the frame that was taken at t_f, in which the particle is

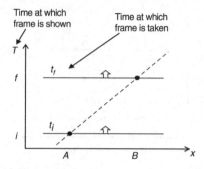

Figure 4.3 A film shown forward

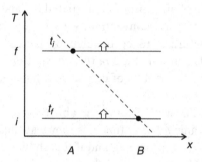

Figure 4.4 A film shown in reverse order

at position B, and ends at time T_f with the frame taken at t_i, in which the particle is in position A (where by our temporal convention $t_i < t_f$ and $T_i < T_f$).

In the reverse-order film we see a particle moving from position B to position A. Since we adhere to the above convention concerning order in space, by which $B > A$, the numerator in the derivative dx/dt should now (that is, in describing the reversed film) be $-dx$. By contrast, the direction *of time*, expressed in the denominator of the derivative, is not reversed: it is still from the start of the *showing* of the film at T_i, to the end of the show at T_f. Therefore, if the velocity seen in the forward film was v, the velocity seen in the reversed film should be $-v$ and the minus sign here is due to the minus sign of dx; there is no reversal of the direction of time here. The sequence of events described in the reversed film satisfies the following equation of motion:

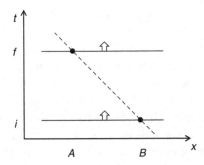

Figure 4.5 (II) A particle moving from B to A

$$A = B + \int_{T_i}^{T_f} \frac{-dx}{dt} \, dt.$$

This sequence of events shown in the reversed film and described by the above equation did not actually take place. However, such a sequence is perfectly consistent with the laws of mechanics; the following can be an actual sequence of states, describing a particle moving from B to A with velocity $-v$ (see Figure 4.5):

$$A = B + \int_{t_i}^{t_f} \frac{-dx}{dt} \, dt. \tag{II}$$

The equation of motion of classical kinematics is such that if evolution (I) is possible, then evolution (II) is possible: if one particular continuous stack of instantaneous time-slices satisfies the law of motion of classical kinematics, then a stack which is similar but is in reverse order also satisfies this law. Of course, which of them (if either) actually obtains depends on the initial state of the system. Since (I) and (II) are equally possible in the theory, the reversed film of the billiard balls seems to us natural; the reversed film is an illustration of the relation between the evolutions (I) and (II) (which is that of velocity reversal).

For this reason, if one is presented with a film of, say, our billiard balls, then one cannot tell whether this film is shown forwards or in reverse. In other words: the two evolutions (I) and (II) are distinguishable since they bring about different sequences of positions; but by merely looking at them it is impossible, as a matter of principle, to know which (if either) is actual and which is counterfactual.

The relation between (I) and (II) is extremely important for understanding some of the difficulties facing the project of underwriting thermodynamics by mechanics. Briefly: if a system that undergoes evolution (I) has its entropy increased, then a reversed evolution (II) is possible by the laws of mechanics, which means that an entropy-decreasing evolution is allowed by the laws of mechanics. Moreover, since to every type (I) evolution there corresponds a type (II) evolution, there are as many entropy-decreasing evolutions as there are entropy-increasing evolutions! This seems to be in clear contradiction of our experience as expressed by thermodynamics, and in particular by the Law of Approach to Equilibrium. This idea, known as Loschmidt's reversibility objection, is one of the central challenges faced by the attempt to underwrite thermodynamics by mechanics. We will discuss this issue in detail in Chapter 7.

4.6 Retrodiction

Consider the following case. We observe a particle at position B at time t_f, and we want to calculate where that particle was earlier, at t_i. Suppose that we know that the velocity of the particle was constant throughout its evolution, and we see that the instantaneous velocity of the particle at t_f is the same as in evolution (I). We naturally conclude that the particle in front of us underwent evolution (I), and we now observe it in the final state of that evolution. If we carry out the right calculations, the outcome should be that at t_i the particle was at position A. The algorithm for doing this, called *retrodiction*, is this:

$$A = B + \int_{t_f}^{t_i} \frac{dx}{dt} dt. \tag{III}$$

Here, the only difference relative to (I) is the order of the integration, which is reversed relative to (I). Equation (III) describes the same evolution described in (I), but whereas in (I) we are given the initial state of position A and velocity v at time t_i, in (III) we are given the final state of position B and velocity v at time t_f. The difference between them is in the order of *inference* only; the retrodiction in (III) is merely a *reconstruction* or a *re-description* of evolution (I), from its last frame to its first frame.

This is an epistemic difference. Equation (III) cannot be interpreted as a physical evolution, since such an evolution would mean that the particle moves from B to A with a velocity that is directed from A to B. For this reason we do not need a separate figure to illustrate (III): it is just as well illustrated by Figure 4.2 which describes (I).

It is of special importance to notice that the direction of time, which forms part of each time-slice, and which is therefore built into the velocity in the derivative dx/dt, remains the same in (III) as it is in (I), and hence the velocity in each of them is the same as in (I). The calculation carries us *backwards* along the sequence of time-slices, but the direction of time built into each of the slices is unchanged. The order of integration and the direction of time are independent of each other.

Similarly, case (IV) is a retrodiction of evolution (II). Given that a particle is in position A and velocity $-v$ at time t_f, and given that its velocity was constant throughout the time interval (t_i, t_f), the algorithm for retrodicting its position at time t_i is the following:

$$B = A + \int_{t_f}^{t_i} \frac{-dx}{dt}\, dt. \tag{IV}$$

The relation between (IV) and (II) is the same as the relation between (III) and (I): while (I) and (II) describe two different possible evolutions, (III) and (IV) are not additional possible evolutions but reconstructions or different descriptions of evolutions (I) and (II), respectively.

Notice that if we read evolution (I) and Equation (IV) as mathematical expressions without being interested in their mechanical interpretation, they are equivalent in the sense that they give the same output for the same input. This is because the double negative sign in Equation (IV) cancels out and may appear to be the same as the positive sign in evolution (I); and similarly with respect to evolution (II) and Equation (III). To see the difference between (I) and (IV) and between (II) and (III) one needs to read into the equations of motion the physical meaning of the steps in the algorithms. The correct physical interpretation is that Equation (IV) is the retrodiction of evolution (II), and Equation (III) is the retrodiction of evolution (I).

The message up to now is this. The four equations (I)–(IV) describe two physical evolutions from two different perspectives: (I) and (II) describe predictions, while (III) and (IV) describe, respectively, their retrodictions. None of the equations described so far involves a reversal of the direction of time, since the direction of time built into each of the time-slices in all four equations (I)–(IV) is the forward direction. Equation (II) is the velocity reversal of (I), not its time reversal; and (III) is the retrodiction of (I), not its time reversal. We now turn to the idea of time reversal in classical mechanics.

4.7 Time reversal and time-reversal invariance

As we saw, a direction of time must be added to each time-slice, if we want instantaneous velocities to be defined uniquely, that is, to have a well-defined direction. But the very requirement to add a direction of time implies that, formally, we can add to the time-slices one direction out of the two possible directions of time, each of which is the *time reversal* of the other. By the term *time reversal* we refer here to a reversal of the time arrow itself, which is built into each of the time-slices, as illustrated in Figure 4.6, in which we reverse the time direction in each slice relative to the direction in Figure 4.1. Classical mechanics is coherent with the idea of a reversed direction of time, and so in order to understand the way in which time functions in this theory, it is instructive to consider the case of a universe with a time direction that is reversed relative to our own.

(Notice that in Figure 4.6 the t axis has not changed relative to Figure 4.1. The direction in the t axis expresses the convention that the numbers attached to each time-slice do not change when the direction of time is reversed; more on this below.) It is a triviality that we always experience one direction of time, namely the direction that we call "from past to future," and we never experience a reversed direction of time. According to classical mechanics our universe is either the one described in Figure 4.1 or the one described in Figure 4.6. The question is, how are these possible universes related to each other? And can we

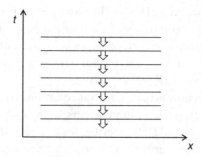

Figure 4.6 Time reversal

distinguish between them? These are the questions we wish to address in this section and the next.

Imagine a universe in which the continuous stack of time-slices is identical to ours in the positions in each of the time-slices, and in the relation of betweenness among the time-slices, and hence also in the instantaneous speed of the particles in each of the time-slices. But this universe differs from ours in the direction of time, which is reversed: the arrow of time which is part of every instantaneous time-slice runs in the direction that we would call future to past. The identity (that is, sameness) of the time-slices in the two universes in the positions of the particles and in the relation of betweenness makes it possible to compare these two universes. Notice that these are two universes, and not the same universe experienced in a different way: since we take the arrow of time to be part of the time-slices that make up a universe, we must have here two distinct stacks of time-slices.

For ease of presentation, imagine an observer – call her Tami – who lives in the time reversed universe. If, for us, the time-slice that we experience at the instant called t_i appears *earlier* than the time-slice we experience at the instant called t_f, then for Tami the time-slice associated with the instant t_i is *later* than the time-slice associated with t_f. The relations of "before" and "after" are reversed when we shift from our universe to Tami's universe. (We do not claim that such a universe exists or that it is compatible with physics in general, only that it is compatible with classical mechanics. We also make no statements about the metaphysical nature of time – whether it flows, or whether events occurring in time "become," etc.)

In what follows we attempt to understand some aspects of the experience of Tami. By the term "experience" we refer to any kind of mental state. However, our approach in this book is physicalist: we take it that mental states are nothing but physical states. We expand on this point in Chapter 5. Right now, the important point in this context is this: since the arrow of time is part of the instantaneous microscopic states of the particles that make up the universe, and since mental states are instantaneous microscopic states of the particles that make up the observer's brain, it turns out that our mental states may well be significantly different from those of Tami, even when both of us go through (as it were) the same time-slice in opposite directions.

We indicate the evolutions experienced by Tami with an asterisk; for instance, whenever in our universe evolution (I) takes place from position A at t_i to position B at t_f, Tami experiences the counterpart evolution (I*),

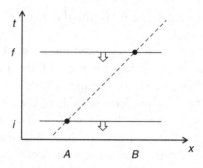

Figure 4.7 (I*) A particle moving from *B* to *A* with a reversed direction of time

from *B* at t_f to *A* at t_i. Evolutions (I) and (I*) are the same with respect to the relation of betweenness, but they differ in the directions of the time arrows; see Figure 4.7.

In order to emphasize the aspects in which evolutions (I) and (I*) are identical and the aspects in which they differ, we adhere to the same conventions as before with respect to ordering positions in space and moments of time: that is, the order of positions in space remains unchanged as we switch between the time-reversed universes, so $B > A$ in both of them. The convention regarding the way in which numbers are associated with moments of time is unchanged as well, that is, if *for us* t_i is earlier than t_f then $t_i < t_f$; and so for Tami $t_i < t_f$ even though t_i is *later* than t_f. We find this convention convenient for the present discussion. Other conventions may have advantages as well as shortcomings; and since this is a convention, the choice is not crucial for the conclusions we wish to draw.

The laws of mechanics, and in particular the law of motion expressed by Equation (1) at the start of Section 4.2 above, hold for Tami since nothing in these laws depends on the choice of a particular direction of time. This is one sense in which classical mechanics is said to be *time-reversal invariant* (TRI).

The reversal of time is expressed in the laws of mechanics by the sign of the denominator of the derivative dx/dt in (1) and (2). Because of the time-reversal invariance (in this sense) of classical mechanics, we are able to write the equations of motion that Tami would write, using the above conventions regarding directions in space and of time. To see how this is done, notice first that in evolution (I) above, which pertains to our forward direction of time (see Figure 4.2), the constant velocity of the particle is given by

$$V_I = \frac{B - A}{t_f - t_i}. \qquad (4)$$

Similarly, in the time-reversed evolution (I*), experienced by Tami, and illustrated in Figure 4.7, the velocity of the particle is given by

$$V_{I*} = \frac{A - B}{t_i - t_f}. \qquad (5)$$

Expression (5) needs some explanation. Regarding the denominator in (5): from Tami's perspective, t_i is later than t_f, and therefore we have in the denominator $t_i - t_f$, which is negative by our convention; and so when we write the equation of motion for evolution (I*) we will write $-dt$ in the denominator of the velocity derivative, to express the fact that it is negative relative to evolution (I).

Regarding the numerator in (5): $A - B$ in the numerator is negative by our spatial convention, which holds also for Tami. Here some subtlety is involved, which needs clarification before we can write the full equation of motion for evolution (I*). The numerator describes the difference not only between any two positions, but between the earlier (for Tami) position B and the later (for Tami) position A, where "earlier" and "later" are determined in parallel to the time difference in the denominator. In this sense, already the numerator in Equations (4) and (5), and hence in the general form (2), presupposes a direction of time. The position differences that go into the velocity expression (2) are not purely geometrical differences, but involve the direction of time from one end of the difference to the other. And so when we write the equation of motion for evolution (I*), we have to write $-dx$, in order to express the fact that its sign is the opposite of the sign of dx in equation (I).

Since both the position difference in the numerator and the time difference in the denominator in (5) and (4) are opposite in sign, it turns out that if the velocity of the particle in evolution (I) as we perceive it is v, then the velocity of the particle as Tami would perceive it in evolution (I*) is v as well. At first glance, this might seem strange, but nothing is really odd here: it is simply a matter of getting used to the convention. Obviously, the directions of the velocities called v in the two universes are opposite in space and in time. Consequently, in order to describe the evolution (I*) in which a particle moves from B at the earlier (for Tami) time t_f to A at the later (for Tami) time t_i, with velocity v, in the reversed time direction, Tami carries out the following calculation:

$$A = B + \int_{t_f}^{t_i} \frac{-dx}{-dt} \, dt. \tag{I*}$$

Having identified the equation (I*) which is the counterpart of (I), let us find the counterpart of Equations (II), (III), and (IV). The velocity reversal (II*) of (I*) is obtained in the same manner as the velocity reversal of (I), namely by replacing the numerator $-dx$ by its opposite dx:

$$B = A + \int_{t_f}^{t_i} \frac{dx}{-dt} \, dt. \tag{II*}$$

The retrodiction (III*) of evolution (I*) is provided in the same manner as above, namely by reversing the order of integration:

$$B = A + \int_{t_i}^{t_f} \frac{-dx}{-dt} \, dt, \tag{III*}$$

and, in a similar way, the retrodiction (IV*) of (II*) is given by

$$A = B + \int_{t_i}^{t_f} \frac{dx}{-dt} \, dt. \tag{IV*}$$

Notice that as before both (III*) and (IV*) are merely reconstructions from end to start of evolutions (I*) and (II*), respectively, and do not describe possible physical evolutions either for us or for Tami, for the same reasons spelled out above for (III) and (IV).

To sum up this section let us emphasize the property of classical kinematics called *time-reversal invariance* (TRI). This property means that the laws of mechanics hold equally in our universe as in Tami's universe, and consequently Equations (1), (2), and (3) apply for both. In our universe this is expressed in evolutions (I)–(IV) while in Tami's universe it is expressed in evolutions (I*)–(IV*). In both universes, each time-slice has an arrow of time as an integral part of it. Without this arrow, velocity is not well-defined since its direction is not determinable uniquely. And since in each of the two universes the time arrows attached to all the time-slices point in the same direction, each trajectory that describes the evolution of a mechanical system has a built-in arrow of time. And so, *although mechanics is time-reversal invariant, each evolution in each of the two universes has a direction of time within every trajectory.*

4.8 Why the time-reversal invariance of classical mechanics matters

The evolutions (I)–(IV) are all of kinds that are familiar to us: they describe processes that we experience or retrodict. By contrast, evolutions (I*)–(IV*) are not part of our experience. Nevertheless they are significant, for they widen the set of counterfactuals that we can consider, and thereby help us understand our own world. In order to see some of their implications consider, again, the billiard game films. The film showing the billiard balls was shot when evolution (I) took place in the world, and when it is shown in a reversed order it appears to us as evolution (II). And so the reversed film illustrates the difference between *experiencing* evolution (I) and *experiencing* evolution (II). However, the reversed film also illustrates Tami's experience of evolution (I*), in the following sense. Suppose that in our world a billiard game takes place from time t_i to time t_f, and in this game ball S undergoes evolution (I) from position A to position B. Assuming that the universe of Tami is identical to ours in all but the direction of time, a game takes place in Tami's world, and in it ball S evolves from B to A during the time interval from t_f to t_i. The sequence of positions in (I*) which Tami experiences is identical to the sequence of positions in the reversed film (II), and in this sense the *internal experiences* of the two observers are identical. So the internal experience of an observer cannot tell this observer whether he is watching a film showing evolution (II) in one time direction or evolution (I*) in the reversed direction of time. And of course a similar argument applies to the indistinguishability of evolutions (I) and (II*). This indistinguishability is another sort of time-reversal invariance in classical kinematics, which we may call *experiential* invariance: this kind of time-reversal invariance is the fact that classical kinematics contains no hint whatsoever that would allow an observer to tell whether he is experiencing (II) or (I*), and equally, whether the case is (I) or (II*). Formally, this property is a result of the mathematical fact that these pairs of evolutions yield the same position output if given the same input.

A comment about the relation between physical (brain) states and mental states is appropriate here. Of course, as we noted above, the instantaneous microscopic states of Tami's brain at each instant are different from our brain states at the same instant, because the arrow of time is part of the instantaneous microscopic states of the particles that make up an observer's brain. If, indeed, experience is invariant under time reversal, this would be a case of multi-realizability. And if this is true, it means that mental states supervene on positions (that is, there can be no difference in mental states without a difference in positions). Since at the present state of science physicalism is a program that cannot be spelled out in detail concerning the way in which physical states make up mental states, we can offer no statement concerning whether this conjecture is true. Here we make the bold assumption that the reversal of the direction of time does not alter the instantaneous mental states. We return to other aspects of multi-realizability in Chapter 5.

Assuming that the theory of classical kinematics is experientially time-reversal invariant, that is, assuming that (II) and (I*) and (I) and (II*) are empirically indistinguishable, the question arises whether it is meaningful to state that we inhabit one universe rather than the other. This question echoes the famous debate between Leibniz and Newton's disciple Clarke about whether space is absolute or relational.[14] This debate is of special importance *if* it is indeed the case that our universe is time-reversal invariant, and in particular whether the difference in the arrows of time brings about a difference in mental states.

Here it is important to remind ourselves that time-reversal invariance is not a trivial property, and not all theories have it.[15] For example, thermodynamics is not time-reversal invariant: if the thermodynamic regularities hold in our universe, then Tami experiences an anti-thermodynamic world: whenever a gas expands in our universe, it contracts in Tami's. This fact is a major obstacle on the way to underwriting thermodynamics by mechanics in *our* universe. For the fact that the universe of Tami is consistent with mechanics means that as far as mechanics is concerned, a universe in which the laws of thermodynamics do not hold even probabilistically is *possible*; and if it is possible, then it may very well be the case that *our* universe is like that. That is, mechanics does not rule out the possibility that our universe will be, in the long run or perhaps starting five minutes from now, anti-thermodynamic. Hence, a universal proof of a theorem of mechanics to the effect that a thermodynamic behavior is highly likely is not possible. This is the most important lesson to be learned from considering Tami's imaginary universe.

Despite the above, it is a fact that the laws of thermodynamics do hold (approximately) in our universe, and the question is why. In other words, what are the special *mechanical* conditions which bring it about that the

---

[14] See Alexander (1956).
[15] Some versions of quantum mechanics are not time-reversal invariant, for example the collapse theory by Ghirardi, Rimini, and Weber (1986).

Law of Approach to Equilibrium and the Second Law hold in our universe? This is a question about the direction of certain processes *in* time rather than the direction *of* time. And this is the central question of classical statistical mechanics that we shall discuss extensively in the following chapters.

5

Macrostates

5.1 The physical nature of macrostates

According to mechanics, the state of the universe at each moment is its microstate, namely the precise positions and momenta of all of its particles. Given this state, together with the parameters of the particles and the constraints and forces that act on them, the entire past and future history of the universe is (in principle) determined. In practice, of course, the full information concerning the state of complex systems, such as thermodynamic ones, is not available to us and, even if it were available, we would not be able to solve the equations of motion of such systems. And so it might seem that the task of describing complex systems, in particular thermodynamic systems, in terms of mechanics, is hopeless. However (as we outlined in Chapter 3), it turns out that we do not need all this information in order to describe or explain the thermodynamic phenomena and regularities; here, partial information suffices. The reason for this is that sets of microstates, those that share certain mechanical properties, exhibit observable regularities that are expressed by the laws of thermodynamics. The amount of information needed to describe the regularities governing these sets is manageable, and for this reason identifying these sets and their regularities is of utmost importance. These sets are called macrostates. Macrostates form the bridge between mechanics and thermodynamics and are the cornerstone of statistical mechanics.

Macrostates are physical, since the properties shared by microstates, which give rise to macrostates, are objective physical properties. Of course, the microstates in the state space can be partitioned into sets according to various criteria, but we will be interested in two of them: thermodynamic distinguishability and thermodynamic *regularity*, as follows.

94

Thermodynamic distinguishability: One kind of physical property according to which sets of microstates can be defined is distinguishability by a given observer: in general, observers are unable to distinguish between individual microstates, but can distinguish between certain sets of microstates; and each distinguishable set of indistinguishable microstates forms a macrostate. In general, different observers can distinguish between different sets, and therefore macrostates are relative to a given observer. This relativity is, however, an *objective* fact about the *physical* correlations between the observer in question and the world that is being observed. We will naturally be especially interested in macrostates relative to *human* observers,[1] which include the sets of microstates that appear to us as the *thermodynamic macrostates*. Human observers happen to be equipped with the capability of perceiving these sets directly and immediately to such an extent that thermodynamic magnitudes may appear to us as elementary properties; it is only the theory of mechanics that tells us that the elementary properties are the mechanical properties, and these appear to us as the thermodynamic properties. For this reason, the sets of microstates that make up the thermodynamic macrostates are of special importance, and have a central role in statistical mechanics.

Thermodynamic regularity: Another kind of physical property of interest, shared by microstates, is the one that gives rise to the thermodynamic regularities. Certain sets of microstates exhibit this particular kind of regularity: all the microstates in these sets appear to satisfy the same laws, described by the theory of thermodynamics.

There are infinitely many other kinds of properties according to which one may partition the microstates of a system into sets: many kinds of observers exist (a fact that we will call the Liberal Stance), and many kinds of regularities are possible. But the above two partitions will be our starting point here, and we now focus on them.

The partition of microstates into thermodynamic macrostates, brought about by the indistinguishability of microstates by (human) observers, explains the way in which (human) observers *perceive* the world; but this kind of partition does not explain the thermodynamic regularities for the obvious reason that it is conceivable, and consistent with mechanics, that

---

[1] And other animals as well.

the sets of microstates that are indistinguishable by human beings (that is, macrostates) would not exhibit any perceived regularity. It is a *contingent fact* about the structure of human beings as observers, that there is a useful degree of overlap between the sets that satisfy the regularities and the sets that correspond to our observation capabilities: because of this overlap we happen – as a matter of contingent fact – to be able to distinguish between the sets of microstates that give rise to the thermo-dynamic regularities, and thus to see these regularities.[2] Of course, the fact that the distinguishable kind of sets fits the thermodynamic kind of sets makes it particularly easy to discover and test the thermodynamic regularities. Still, these two ways of partitioning the microstates into sets have different physical causes, and so are conceptually different, and the distinction between them is a central point in understanding statistical mechanics.

It is important to realize what exactly is the explanatory role played by the partitioning of the state space into thermodynamic macrostates. Since macrostates are sets of microstates that are indistinguishable by an obser-ver, and since the actual state of the universe at any moment is a single microstate, it turns out that the statement that a system is in some macrostate must involve some sort of *ignorance* on the part of the obser-ver, as to which of the microstates in the macrostate is actual. Now, this *ignorance clearly does no work in explaining the thermodynamic regularities* which result from the way in which sets of microstates happen to behave. But this ignorance does explain why we happen to see these specific regularities – namely, the ones describable in terms of thermodynamic macrostates – rather than any other regularities that may exist in the world but cannot be perceived by human observers. It is quite conceivable that there are other sets of microstates in the universe (uncorrelated with the thermodynamic sets) that satisfy certain regularities, but since we – as observers – are not correlated with these sets, since (in other words) these sets do not form macrostates for us, we do not see these regularities. And so, we need both the above notion of ignorance, which brings about macrostates, and the factual regularities of certain sets of microstates in order to explain the *thermodynamic* regularities of *thermodynamic* macro-states on the basis of mechanics.

It is also crucial to realize, as we will now show, that the above notion of ignorance is objective and physical: it involves no subjectivity.

---

[2] We propose no hypotheses as to the origin of this coincidence.

In this chapter we show that this so-called ignorance with respect to the distinction between the actual microstate and the counterfactual microstates is a purely objective mechanical fact. As we said in Chapters 1 and 3, classical mechanics is *a theory without an observer*: notions such as observer, ignorance, and so on have no primitive role in it; and while such terms are sometimes convenient they have no fundamental significance. In this chapter we will describe their mechanical counterparts.

An important example of the interplay between thermodynamic distinguishability and thermodynamic regularities, which is the heart of statistical mechanics, is the first known macrostate, namely the so-called *Maxwell–Boltzmann distribution* (see Section 5.7). Maxwell and Boltzmann characterized a set of microstates all of which share the same distribution of velocities among the particles that make up an ideal gas, and showed that a gas in *any* of these microstates will appear to us to be in the same state of thermodynamic equilibrium. They discovered that certain observable regularities of the behavior of the gas, known from thermodynamics, depend only on the distribution of the velocities and are independent of the particular microstate in this set of microstates. This independence means that the gas would appear to have the same thermodynamic properties as long as its microstate belongs to the macrostate called the Maxwell–Boltzmann distribution. It is instructive to notice that the Maxwell–Boltzmann macrostate is both a set of microstates that are indistinguishable by us (human beings) as observers (as we explain in Section 5.7), and a set that satisfies the thermodynamic regularities. It is, therefore, an important example of the interrelations between these two conceptually different ways of partitioning microstates into sets that form macrostates. Of course, it is only in retrospect that we can say that this was the first time in which a physical state was described in terms that we now call a macrostate. But although this first appearance of the notion of a macrostate in physics was to a large extent implicit, it had a great influence on the development of statistical mechanics. We discuss this important example later in this chapter.

5.2 How do macrostates come about?

In light of the above discussion, two major questions ought to be asked. The first is: how do the thermodynamic regularities come about? Do they follow from the laws of mechanics, that is, are they theorems provable from the principles of mechanics? We address this question in Chapter 7.

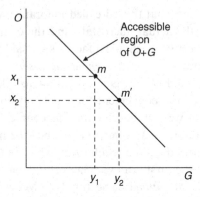

Figure 5.1 Accessible region with one-to-one correlation between O and G

The second is this: how does the thermodynamic distinguishability come about – as an objective feature of the physical interaction between the systems that make up the observer and observed? This is the question we focus on now.

Consider a toy universe consisting of particles that obey the laws of classical mechanics. Now divide this universe into two parts, and call one of these parts O and the other part G (we give these parts no particular properties or role at this stage). The microstate of the composite system $O + G$ is associated with a point in a multi-dimensional state space, in which each dimension stands for one of the mechanical degrees of freedom of $O + G$. The limitations on $O + G$ (such as their total energy), the external constraints imposed on them (such as the positions available to each of them), and their interactions with each other (in virtue of their parameters such as mass or electrical charge) together determine a region in their state space which contains all the microstates that $O + G$ can occupy, namely the accessible region of $O + G$. All the points in the accessible region are counterfactually possible for $O + G$ in the sense that they are consistent with their parameters, limitations, external constraints, and internal interactions.

Figures 5.1 and 5.2 depict two examples of accessible regions of $O + G$, where each axis represents a set of degrees of freedom. In Figure 5.1 the accessible region is a section of a straight line in the OG plane. Suppose that $O + G$ is in the microstate m. The projection of m on the axis of O is x_1 and on the axis of G is y_1. Here the structure of the accessible region implies that there is a one-to-one correlation between the microstates of O and the microstates of G: for any microstate of O there corresponds a unique microstate of G, and *vice versa*. As a result, given the accessible

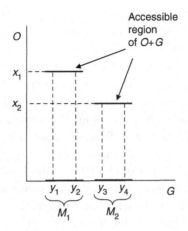

Figure 5.2 Accessible region with one-to-many correlation between O and G

region of $O + G$, one can deduce the microstate of G from the microstate of O, and *vice versa*.

To illustrate the discussion, think for example of G as a particle that can be in one of several positions, and of O as an automaton that points its hand in the direction of G's position. Here one may call O an *observer* and G the *observed system*. This is only for the sake of being concrete and figurative; nothing in this terminology implies anything about the physical properties of O and G, for their properties are only those described here in terms of their joint state space. In these terms, given the accessible region as described in Figure 5.1, one may say that O has a microscopic *resolution power* with respect to the state of G, since each microstate of G is correlated with a different microstate in O, owing to the structure of the accessible region. Of course, in this case the opposite holds as well: each microstate of O is correlated with a different microstate in G. We discuss physicalism in Section 5.8.

In Figure 5.2 the accessible region of $O + G$ consists of the two thick line segments in the OG plane. The correlations between O and G in Figure 5.2 are one-to-many: each microstate of O is correlated with several (in this case, a continuum of) microstates of G. For example, the microstate x_1 of O is correlated with both microstates y_1 and y_2 of G; and the microstate x_2 of O is correlated with microstates y_3 and y_4 of G. Owing to this structure of the accessible region, one cannot infer the microstate of G from the microstate of O. In this sense, O cannot distinguish between the microstates y_1 and y_2 of G, nor between the microstates y_3 and y_4 of G. Microstates y_1 and y_2 of G belong to the same equivalence class relative to the microstate x_1 of O, and similarly for y_3 and y_4 relative to x_2. In this case we shall say that the pair of points y_1 and y_2 belongs to one *macrostate* of G, which we denote by $[M_1]$ and the pair of points y_3 and y_4 belong to a different *macrostate* of G, which we denote by $[M_2]$. As we said in section 3.5, we denote regions that form macrostates by "$[\ldots]$".[3]

In our account of macrostates in mechanical terms, a macrostate is a set of microstates of G all of which correspond to a single microstate of O. In such an account, macrostates reflect the resolution power of O with respect to G, and therefore these macrostates are *of G relative to O* (they are not macrostates of O, nor of $O + G$). For this reason the macrostates $[M_1]$ and $[M_2]$ of G in Figure 5.2 are regions (represented by thick line segments) *on* the horizontal axis G, and not in the OG plane. The line segments in the OG plane are the accessible region of $O + G$, while the thick line sections on the G axis are the macrostates of G. In other words, the interaction with O brings about a partition of the state space of G into regions corresponding to the macrostates of G.[4]

It may happen that the correlations between the microstates of O and G are many-to-many. In Figure 5.3, for example, the region O_1 on the O axis is a set of microstates that share some well-defined mechanical property, for example a range of positions, or average velocity, and similarly for the region O_2 on the O axis. The correlations between each of the microstates in O_1 (or O_2) and the microstates in $[M_1]$ (or $[M_2]$, respectively) are the same as described in Figure 5.2. For this reason $[M_1]$ and $[M_2]$ are macrostates of G.

It is important to distinguish between three aspects of Figure 5.3. (i) The rectangles on the OG plane are the accessible region of $O + G$; (ii) the two thick line segments on the G axis are the macrostates $[M_1]$ and $[M_2]$ of G relative to O; and (iii) each of the two thick line segments O_1 and O_2 on the O axis is a set of microstates of O each of which is associated with the corresponding macrostate of G.

---

[3] Two important comments on the model in Figure 5.2: (i) Accessible regions are generally *continuous*, and so you may think of the line segments in the OG plane as being either *incomplete*, for example they may be projections of a space with additional dimensions in which these line segments are connected; or *imprecise*, for example you may bring them close together in your imagination along the O axis until their ends touch. We chose the illustration as it is in Figure 5.2 (and elsewhere) for clarity of presentation. (ii) The particular structure of the accessible region in Figure 5.2 induces an asymmetry between O and G: while one cannot infer the microstate of O from the microstate of G, one *can* infer the microstate of G from the microstate of O. This asymmetry is purely physical, and is an outcome of the particular constraints and limitations on the universe consisting of $O + G$.

[4] See also Earman (2006) concerning the idea that there is one partition of the state space to thermodynamic macrostates that is the maximally fine-grained one.

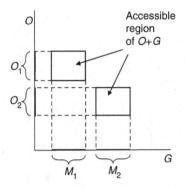

Figure 5.3 Accessible region with many-to-many correlation between O and G

There are good reasons to believe that the interaction between human beings (as observers) and our environment is of the kind described by Figure 5.3 rather than Figures 5.2 or 5.1 (see Section 5.8). Notice that since O and G are purely mechanical systems, there is no fundamental difference in their physical status; and indeed, Figure 5.3 expresses a certain similarity in their status in this respect. The only asymmetry in the figure is the way we interpret the systems O and G, and perhaps the details of the structure of O that make it an observer to which *experience* can be ascribed (more on this in Section 5.8). Indeed, since O is a mechanical system just like G, it too can be assigned macrostates. But since macrostates are sets of microstates that are indistinguishable *by a given observer*, in order to assign macrostates to O one has to introduce another physical system O' which interacts with O in such a way that the correlations between O and O' bring about a partition of the state space of O into equivalence classes *relative* to O'. For example, G in Figure 5.3 can be the observer of O, provided that G has the appropriate physical structure. In this case, the regions O_1 and O_2 in Figure 5.3 would be the macrostates of O relative to G, and the regions M_1 and M_2 would contain the microstates associated with G's experience.

Since the correlations between O and G are one-to-many (or many-to-many), in order to have a mechanical theory of macrostates one must be able to characterize the aspect of the microstates of G in virtue of which all of them get correlated with the same microstates of O in purely mechanical terms. In other words, the characterization of the macrostates of G must be given in *intensional* terms, namely in terms of physical properties.[5] If the

---

[5] As opposed to an extensional description of a set, that is, by specifying its members.

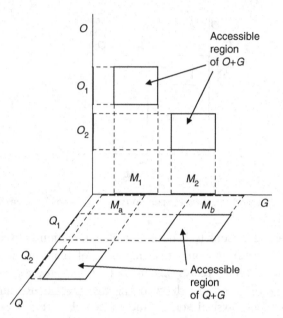

Figure 5.4 Macrostates (M_1 and M_2) of G relative to O and macrostates (M_a and M_b) of G relative to Q

characterization were to be only extensional, we would not have a manageable underwriting of thermodynamics by mechanics.

An important aspect of the macrostates of G, in this account, is that while they are relative to O, this relativity is purely objective, since it is determined by the structure of the accessible region which is, in turn, determined by the limitations on the energy and positions available to our toy universe. To see this point, suppose that our toy universe were divided into three systems: G, O, and Q, where if O is one observer then Q is another observer. Figure 5.4 depicts the projections of the total accessible region of $O+Q+G$ on the OG plane and on the QG plane, thus showing the relations between O and G and between Q and G. The relations between O and G in our example are exactly like their relations in the $O+G$ universe, depicted in Figure 5.3, and bring about the macrostates $[M_1]$ and $[M_2]$ of G relative to O. The relations between Q and G, exhibited by the projection of the total accessible region of $O+Q+G$ on the QG plane, would look qualitatively the same as the relations between O and G in Figure 5.3, namely they would consist of two rectangles in the QG plane. But the details of this projected accessible regions are different: the regions along the G axis, that make up the macrostates of G relative to Q, are the regions $[M_a]$ and $[M_b]$, which are *not* the same as the regions along the G axis which make up the macrostates $[M_1]$ and $[M_2]$ of G relative to O. The macrostates $[M_1]$ and $[M_2]$ of G relative to O and the macrostates $[M_a]$ and $[M_b]$ of G relative to Q are both determined by the accessible region of the universe that consists of $O + Q + G$, and this accessible region is determined by physical properties of this universe. It is in this sense that although macrostates are relative to an observer, *this relativity is objective and physical*.

This is how mechanics can explain the fact that our experience can be described in terms of macrostates, despite the fact that at the fundamental level the actual state of the universe is a microstate.

A very important remark

It is important to point out at this stage that there is an apparent difficulty concerning the notion of macrostates that will be solved later on in the book. Although the notion of macrostates is the central one in statistical mechanics, since it explains the meaning of the thermo-dynamic magnitudes and gives rise to the probabilistic versions of the laws of thermodynamics (as we explain in Chapter 6), the actual precise identification of sets of microstates that give rise to thermo-dynamic macrostates is often very difficult: owing to the immense complexity of thermodynamic systems and the details of our inter-actions with such systems, it may in general be practically *unfeasible* to calculate the regions of the state space corresponding to the thermo-dynamic magnitudes. For example, below we describe how Boltzmann partitioned the state space of an ideal gas into macrostates, but even this successful result is based on empirical conjectures (concerning the partition of the µ-space into cells) that are not detailed enough to provide an accurate identification of the state-space regions that cor-respond to macrostates. However, this difficulty is overcome in statis-tical mechanics: in Chapter 11 we will show a way to circumvent the need to calculate macrostates in some interesting thermodynamic cases. We shall refer to this remark later on as the *very important remark from Chapter 5*.

5.2.1 Measuring devices and macrostates

In the discussion so far we have described a case where the observer O observes the system G directly. But observers often measure the state of the observed system indirectly, by using measuring devices. Let us

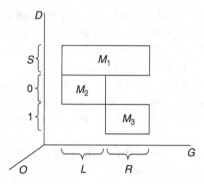

Figure 5.5 *O* observes *G* using the measuring device *D*

describe a situation in which *O* measures some quantity of *G* with a measuring device *D*. For illustration, consider the case in which *G* is a particle which can be in either the left-hand side or the right-hand side of a box, and *D* is a device which has a pointer that indicates the region marked *S* on its dial when *D* is in its pre-measurement Ready state, the region marked 1 on its dial when *G* is on the right-hand side of the box, and the region marked 0 when *G* is on the left-hand side of the box. The observer *O* looks at the state of *D* and infers from it the location of *G*. For simplicity of presentation, let us assume that in this case *O* could also, in principle, measure *G*'s location directly, without using *D*.

The state space of this case is illustrated in Figure 5.5, where the vertical axis *D* stands for the measuring device. The external constraints upon *O, G*, and *D* and the internal interactions between them determine their common accessible region. However, it is sometimes convenient to consider the projection of this accessible region on *O* and *G* only, or on *O* and *D* only, in order to learn about the correlations between these pairs. For example, the correlations between *O* and *G* are such that the macrostates of *G* (relative to *O*) are the regions [*L*] (for left) and [*R*] (for right) on the *G* axis, and the correlations between *O* and *D* are such that the macrostates of *D* (relative to *O*) are the regions [*S*], [0], and [1] on the *D* axis. (In the case where *O* is unable to observe *G* directly, the macrostates of *G* relative to *O* would depend on the interaction between *O* and *D* as well as the interaction between *D* and *G*.) For simplicity we do not indicate in the figure the regions on the *O* axis. Taking the macrostates of *G* and the macrostates of *D* together, one may talk about the macrostates of *D* + *G* relative to *O*. In Figure 5.5 these macrostates are denoted by the two-dimensional rectangles [*M*_1], [*M*_2], and [*M*_3].

The dynamics of the measurement interaction maps the points in the left half of $[M_1]$ to the points in $[M_2]$ and the points in the right half of $[M_1]$ to the points in $[M_3]$. As before, the macrostates $[M_1]$, $[M_2]$, and $[M_3]$ are equivalence classes defined by the many-to-many correlations between the microstates of O and the microstates of $D + G$, precisely as explained above.

5.3 Explaining thermodynamics with macrostates

As we said earlier, there are two conceptually different kinds of partitions of microstates in the state space to sets: one arising from the observation capabilities of an observer and the other arising from the thermodynamic regularities governing the evolution of sets of states; and it is both the difference between these two kinds of sets, and the way in which they are linked, which explains the observed thermodynamic regularities.

As we saw in Chapter 2, there are two kinds of thermodynamic regularities. One kind holds for properties of thermodynamic systems at a given moment of time, in particular when the system is in a state of equilibrium, such as the ideal gas law $PV = nRT$ (which we address in this section). Another kind of regularities are those that hold as a system evolves in time, such as the Law of Approach to Equilibrium and the Second Law of Thermodynamics (which we address in the next section).

The regularities that hold in equilibrium characterize the way in which human observers happen to be correlated with other systems. The fact that thermodynamic behavior depends on certain aspects of the particular microstate of the system is expressed by the one-to-many (or many-to-many) correlations between O and G. Statistical mechanics attempts to identify and express the aspects shared by many microstates, which give rise to similar thermodynamic behavior, in terms of intensional mechanical definitions of sets of microstates. These shared aspects of microstates are directly perceived by human observers and are therefore the basis for the formation of the thermodynamic macrostates. It is a contingent fact about the world that the set of microstates in the accessible region is such that certain regularities hold between these aspects of microstates for systems that are in the equilibrium macrostate (the notion of equilibrium macrostate is further described in Chapter 7). An important example here is the ideal gas law, $PV = nRT$. Let us expand a bit on the mechanical meaning of this law and the terms in it.

In identifying thermodynamic magnitudes with mechanical macrostates, it is important to remember that the standard thermodynamic magnitudes are defined for equilibrium states only, and therefore any attempt to construct a mechanical underwriting of thermodynamics ought to be such that the thermodynamic laws will be reconstructed for certain macrostates, those that would be identified as equilibrium macrostates. Consider, for example, the ideal gas law of thermodynamics, which says that ideal gases in equilibrium satisfy the equation $PV = nRT$. The regularities expressed by this law are objective features of the world, which pertain to the thermodynamic magnitudes that we experience: the volume V, pressure P, and temperature T of the gas. In mechanical terms, volume is the spread in position of the particles of the gas. Because of the structure of the accessible region of $O + G$, there is a one-to-many[6] correlation between the microstates of O and the position degrees of freedom of G, so that O cannot distinguish between the position degrees of freedom of certain microstates of G, and consequently we say that these microstates correspond to the same spatial volume of the gas. In other words, we group these microstates – that are indistinguishable in their spatial spread – within the same volume macrostate. Similarly, the temperature of an ideal gas is characterized in mechanical terms as the average kinetic energy of the particles, and thus expresses one-to-many (or many-to-many) correlations between O and G's velocity degrees of freedom; and the pressure, which is characterized as the number of collisions between the particles of the gas and the container's walls per unit area per unit time, is expressed as correlations between O and G which depend both on position and velocity. In this way, each of the three ways of partitioning the state space of G into macrostates is determined by the interaction between O and G. In each such partitioning there is one macrostate that corresponds to the value of the corresponding magnitude in thermodynamic equilibrium. The regularity expressed by the ideal gas law is thus represented as follows: the three equilibrium macrostates in the three ways of partitioning stand in the relation given by the ideal gas law, and whenever the *actual microstate* of an ideal gas belongs to all three equilibrium macrostates, the gas satisfies the relation $PV = nRT$.

5.4 The dynamics of macrostates

We now turn to the second kind of thermodynamic regularities, namely those that hold for the evolution of thermodynamic systems *in time*,

---

[6] We discuss multi-realizability below.

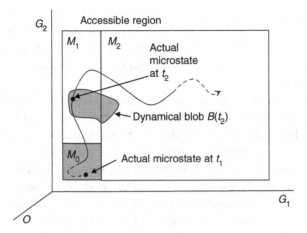

Figure 5.6 Macrostates and dynamical blobs

through different macrostates. It turns out that given the thermodynamic macrostates, the evolution of systems through these macrostates satisfies certain regularities over time. In mechanical terms this means that trajectories that start out in certain macrostates evolve through other macrostates in ways that give rise to such regularities. An important example of such regularity is the Law of Approach to Equilibrium, which we analyze in detail in Chapter 7. The question of whether this kind of regularity is a theorem of mechanics is the holy grail of statistical mechanics. It turns out, as we shall see, that the answer to this question is in the negative.

A mechanical description of a *macroscopic evolution* over time must involve both the notion of a macrostate and that of a dynamical blob (see Chapter 3). In Figure 5.6 we describe a system that starts out at time t_1 in macrostate $[M_0]$. The microstates of $[M_0]$ evolve so that at time t_2 they arrive at the dynamical blob $B(t_2)$. In Figure 5.6 the macrostates of the degree of freedom G_1, which should be thought of as regions along the axis G_1, are determined by the correlations between O and G_1; and similarly for the macrostates of the degree of freedom G_2. And so the macrostates $[M_0]$, $[M_1]$, and $[M_2]$ in Figure 5.6 express the correlations between O and G_1 and between O and G_2, together.

By contrast, the dynamical blob $B(t)$ describes the trajectories of the entire universe which is, in our case, $O + G_1 + G_2$. In other words, $B(t)$ in Figure 5.6 should have three dimensions. It evolves not only in the plane spanned by G_1 and G_2, but also in the third dimension, of O; it is this evolution that brings about changes in the microstate of O. For ease of presentation, however, in Figure 5.6 we describe only the projection of the three-dimensional blob of $O + G_1 + G_2$ on the two-dimensional plane of $G_1 + G_2$; this is only a matter of convenience of illustration, which we will use throughout this book. The reader should always bear in mind that $B(t)$ actually resides in the state space that includes the degrees of freedom of the observer O.

The full dynamical blob, including O, must satisfy Liouville's theorem, introduced in Chapter 3. By saying that this theorem should be satisfied, we mean that it should be satisfied in the full state space, including O, although the illustrations will often omit the degree of freedom of O, and therefore appear to express only the projections of the blobs onto the observed degrees of freedom ($G_1 + G_2$ in this example). The macroscopic (thermodynamic) regularities over time are described by the interplay between the time-evolved blob $B(t)$ and the macrostates. This interplay takes place in the projected state space onto the observed degrees of freedom. In order to describe in full the interplay between macrostates and blobs which gives rise to the thermodynamic regularities, we still need a notion of *probability* in statistical mechanics. We develop this notion in Chapter 6, and then in Chapter 7 give the complete statistical mechanical counterparts of the thermodynamic regularities.

5.5 The physical origin of thermodynamic macrostates

There are, in general, two ways to associate the mechanical properties of particles with thermodynamic magnitudes. One way can be thought of as a *conceptual* linkage. For example, it is natural and almost self-explanatory to identify the thermodynamic magnitude of "volume" with the mechanical property of "spread of positions." On the other hand, the identification of the thermodynamic magnitude of "temperature" with the mechanical property of "velocity distribution" of some sort, e.g. average kinetic energy in the case of an ideal gas, is less trivial, and it is harder to see why it should be a consequence of conceptual consider-ations. In this case it seems that the association of temperature with velocities is a generalization from experience. For instance, think of Joule's experiment, described in Chapter 2. The kinetic energy of the propeller is transferred to the water molecules, as the wings of the propeller collide with these molecules. At the same time we experience a rise in the temperature of the water, and we conclude that this rise in temperature is the way in which we sense the increase in the molecular velocities. It is only in retrospect, and following similar experiences, rather than abstract conceptual considerations, that we identify temperature with distribution of molecular velocities.

To see that the partition of the state space into thermodynamic macrostates is discovered in experience, think for example about a particle in a box of width $L+R$, where L and R denote the left-hand and right-hand side of the box, respectively. Suppose that the observer O does not look at the box in order to find out the location of the particle in the box. In this case, since the only correlation between O's brain state and the location of the particle is that O merely knows the particle is somewhere in the box, one can say that the macrostate of the particle is $[L+R]$. But of course if O were to look into the box the macrostate of the particle could be $[L]$ or $[R]$ or some other macrostate corresponding to a smaller volume of the box. In this latter case the correlation between O's brain state and the location of the particle would correspond to such finer-grained macrostates (see also Chapter 9). By contrast, human observers are correlated with the velocity degree of freedom in a different way. It is a fact about our physical structure that we are correlated with the velocity degrees of freedom in a way that brings about a partition that results in thermodynamic notions such as temperature and pressure.[7] In this sense, the thermodynamic partition along the velocity degrees of freedom is *maximally fine-grained*. This fact is contingent in the sense that it depends on the special details of our physical make up and on the special details of the way in which we interact with the environment. There is nothing *a priori* necessary about this make up or this interaction: other kinds of observers, for whom different sets of microstates are indistinguishable, are conceivable, and consistent with mechanics.

Nevertheless, it seems that there is something special about the set of thermodynamic macrostates and the regularities that they exhibit, namely, the regularity described by the laws of thermodynamics, such as the Law of Approach to Equilibrium and the Second Law of Thermodynamics. It is not trivial, and may even seem to be almost a miracle,[8] that the set of macrostates that we happen to experience exhibits such a regularity. However, as we said, this regularity is itself contingent, and not necessary, in the sense that

---

[7] This last statement may look like the assumption of molecular chaos (see Uffink 2007), which is notoriously problematic in statistical mechanics, but here we do not argue that this assumption holds universally, only sometimes for some observers. An example in which molecular chaos is violated is Maxwell's Demon.

[8] See for example Albert (2000, pp. 23–24).

nothing in *mechanics* entails that the thermodynamic macrostates must exhibit such regularity.[9]

Is there an explanation for the fact that the set of macrostates that we happen to experience, happens to exhibit regularity? Although there is no mechanical explanation, there may perhaps be another kind of explanation for it. One possible explanation for this contingent fact (an incomplete explanation, though) is *evolutionary*. For given that the sets of microstates that make up the thermodynamic macro-states *objectively* behave in some regular way, there is a clear advantage in being able to exploit this fact. If agent O comes to evolve in such a way that its correlations with the microstates of G give rise to macrostates of G that exhibit a robust regularity, it will have survival advantages over other agents. One advantage is that O would be able to store information about the world in a more condensed way. Another is that O would be able to formulate successful inductive generalizations: if the correlations between the microstates of O and the microstates of G were microscopic (as in Figure 5.1), no two microstates of the universe would look the same, and therefore O would not be able to make inductive inferences. Statistical mechanics brings in *the notion of sameness* of microstates, to which mechanics is blind, and in this sense it is more informative than mechanics. Of course, by saying that an observer who experiences the world in terms of thermodynamic macrostates has a survival advantage, we do not mean to say that our partition to macrostates is necessarily the opti-mal partition: there may be other agents that carve up the state space into radically different macrostates, which exhibit a regularity that is radically different from the laws of thermodynamics. It may be that the macrostates experienced by those agents have a better fit, a stronger advantage, than ours. But as is well known, evolution does not always result in the best conceivable outcome; this explanation of the physical origin of macrostates is as good as evolutionary explan-ations can be.

---

[9] A different theory of macrostates in statistical mechanics has been proposed by Shalizi and Moore (2003). Their conception of a thermodynamic macrostate is based on the idea that the thermodynamic regularity is the definitive or essential feature of what macrostates are. This idea seems to us misguided for two reasons. First, thermodynamic regularity and thermodynamic indistinguishability are two independent concepts, and only the interplay between them may explain thermodynamics. Second, Shalizi and Moore characterize the thermodynamic regularity by the requirement of a Markov process. This requirement is not necessary or sufficient as a mark of thermodynamic regularity.

5.6 Boltzmann's macrostates

The idea that macrostates are physically objective in the sense that they express one-to-many (or many-to-many) correlations originates in Boltzmann's ideas concerning macrostates, to which we now turn. We shall put Boltzmann's ideas in terms that fit our general account of physical macrostates; that is, we shall describe Boltzmann's macrostates as sets of microstates that are indistinguishable by a given observer, owing to the correlations between the observer and the observed, as we explained above.[10]

Boltzmann[11] starts his construction from the representation of the microstate of *one* molecule of an ideal gas: a single free molecule's microstate is represented, as usual, by a point in the state space of that molecule, which is a six-dimensional space corresponding to its three position and three momentum degrees of freedom; this state space is called[12] the μ-space (or molecular space). Next, Boltzmann turns to account for the fact that observers have limited observation capabilities, and therefore are unable to discern the precise microstate of each gas molecule. In other words, Boltzmann wants to represent the fact that for any (finitely capable) observer O, there are sets of microstates of the gas in the μ-space between which this observer cannot distinguish. To do so, Boltzmann proposed to partition the μ-space into cells such that the microstates of a molecule inside a given cell are *indistinguishable*.[13] This idea is illustrated in Figure 5.7. Suppose, for example, that the microstate of molecule A evolves from microstate s_1 in cell c_1 to another microstate s_2, which is in the same cell c_1. Although the final state of affairs is different from the initial one, the two are deemed indistinguishable; O would say that no evolution took place in this case.

It is important to emphasize that for Boltzmann this kind of indistinguishability between the microstates of a molecule does not yet make a macrostate. Given our above account of macrostates, one might have

---

[10] Our account is based on that of Ehrenfest and Ehrenfest (1912). See also Uffink (2004).
[11] As understood by Ehrenfest and Ehrenfest (1912).
[12] By Ehrenfest and Ehrenfest (1912).
[13] Boltzmann set only vague limitations on the size of the cells in the μ-space: on the one hand, he stipulated that they must be large enough to be able to contain the microstates of several gas molecules at a time (given the density of the gas); and on the other hand, they must be small enough that the microstates within a cell are indistinguishable. The idea of an upper limit on the cell's size implies an observer O, whose presence makes the notion of indistinguishability meaningful.

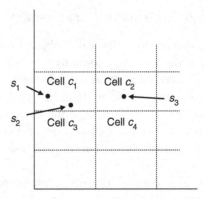

Figure 5.7 Boltzmann's macrostates in μ-space

thought that the set of molecular microstates that are indistinguishable owing to being in the same cell make up a macrostate. But in the context of Boltzmann's account this would be a mistake; Boltzmannian macrostates come later, as we shall see.

Next, Boltzmann turns to account for the implications of this indistinguishability between the microstates of a *single* molecule for the description of the state of an ideal gas consisting of N such molecules. To do so, Boltzmann describes the microstate of *all* the N gas molecules by a cloud of N points in the μ-space. Now consider the way in which the observer O would describe the microstate of the gas, given O's limited observation capabilities represented by the partitioning of the μ-space into cells; see Figure 5.7. Suppose first that, as before, the gas molecules change their microstate in such a way that no molecule leaves its cell; for instance, molecule A evolves from microstate s_1 to s_2 in the same cell c_1, and similarly for the other molecules. In this case, as we said, the observer O would detect no change in the state of the gas; and we shall say that the gas as a whole remains in the same arrangement. By the term "same arrangement," we mean that the same molecules remain within the same cells, although each molecule may change its microstate within its cell. Each arrangement is compatible with infinitely many microstates, since each cell contains a continuum of points.

Now, if molecule A of the gas is in microstate s_1 in cell c_1 and molecule B of the gas is in microstate s_2 of the same cell c_1, and if these two molecules are identical,[14] then a *permutation* between them – such that

---

[14] In the classical context the molecules differ in their haecceity – that is, their labels – despite their indistinguishability.

molecule A is now in microstate s_2 and molecule B is in microstate s_1 – belongs to the *same arrangement*.[15] The observer O would be unable to distinguish between the two permuted states of the identical molecules, owing to O's limited observation capabilities.

Here, one encounters a problem of representation, since permutations between two molecules cannot be represented in the μ-space: whether A is in s_1 and B is in s_2 or vice versa, both cases would look the same – they would look like the two points in cell c_1 in Figure 5.7. In order to represent molecular permutations we need to use the state space, which in the context of Boltzmann's argument is called the Γ-space (or gas space). Here, each molecule is represented by its own six dimensions in the state space, and two permuted molecular states are represented by two different points.

While the observer O cannot distinguish between two permuted states of identical molecules that are within the same cell, it can distinguish between two permuted states that belong to different cells. Consider the following permutation between the identical molecules A and B. Before the permutation, molecule A is in microstate s_1 in cell c_1 and molecule B is in microstate s_3 in cell c_2 (call this state AB); after the permutation A is in s_3 and B is in s_1 (call this state BA). This permutation results in a change of *arrangement*. But it does not change the *distribution* of the molecules, that is, the number of molecules in each cell, since it is the same in the two arrangements AB and BA; see Figure 5.7. Each distribution is compatible with many arrangements. Boltzmann argued that arrangements belonging to the same distribution are indistinguishable, since the observer O is unable to distinguish between arrangements such as AB and BA. Boltzmann's definition of a macrostate of the gas is the set of all the microstates that belong to all the arrangements in a given distribution: all these microstates are indistinguishable by the observer O. This definition is illustrated in Figure 5.8. In the Γ-space, unlike the μ-space, AB and BA are described by two different points. More precisely, since an arrangement consists of the *continuous* set of microstates that are within a given cell, each of the arrangements AB and BA is represented by a *region* in the Γ-space. A distribution is the union of such regions. The region that stands for arrangement AB contains microstates that differ from each other by the positions and momenta within the same cells. The observer O is unable to distinguish between these microstates. The region that

---

[15] For a discussion of this point, see Albert (2000, Ch. 3).

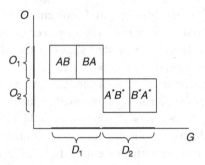

Figure 5.8 Boltzmann's macrostates in Γ-space

stands for arrangement *BA* contains microstates that differ from those of *AB* by permutations of molecules across cells. The union of *AB* and *BA* makes up the distribution D_1, which is a macrostate of the gas *G*. Similarly, the union of the arrangements $A*B*$ and $B*A*$ makes up the distribution D_2, which is another macrostate of the gas *G*. All the microstates in D_1 are correlated with the set O_1 of microstates of the observer *O* which give rise to the same experience. And the situation is similar for the microstates in D_2.[16]

Boltzmann, it seems to us, attempted to account for the partition of the state space into macrostates by focusing on two physical causes for the correlations between *O* and *G*: the indistinguishability of microstates belonging to the same cells in the μ-space, and the indistinguishability between microstates that differ in permutations of particles across different cells.

Boltzmann established the empirical significance of his construction of macrostates in the following way. By counting the number of

---

[16] One of Boltzmann's central aims in his construction was to characterize different macrostates according to their *sizes* in the state space of a thermodynamic system. To get a notion of size, Boltzmann counts each arrangement in the μ-space as a single state, and then counts how many arrangements are compatible with each distribution. However, since each arrangement in the μ-space is compatible with a continuum of microstates, counting each arrangement as a single state essentially presupposes the Lebesgue measure (or measures that are absolutely continuous with the Lebesgue measure). On the basis of this assumption, Boltzmann gives a straightforward combinatorial argument according to which the macrostate corresponding to thermodynamic equilibrium of an ideal gas (i.e. the Maxwell–Boltzmann distribution) is far larger than any other macrostate of the gas *provided* that one measures the size of the cells in the μ-space by the Lebesgue measure. But in order for this argument to have a physical significance, one must find some criterion for preferring the Lebesgue measure over other measures. Boltzmann's criterion seems to have been that the Lebesgue measure corresponds to probability via the idea that a set of microstates that has a large measure also has high probability. But this claim is unjustified, as we argue in Chapter 8 and elsewhere in this book.

arrangements in each distribution using combinatorial calculations, and assuming that all the cells in the μ-space have the same (Lebesgue) measure, he was able to identify and characterize the macrostate that he found to be the largest, as the one corresponding to the so-called Maxwell–Boltzmann energy distribution (see more details about this macrostate in Section 5.7). As we have already mentioned, Maxwell and Boltzmann were able to derive some predictions from this distribution, concerning the behavior in equilibrium, and their predictions have been empirically successful. By showing that the largest macrostate in the above combinatorial sense is the Maxwell–Boltzmann energy distribution macrostate, Boltzmann made the linkage between his notion of macro-states and experience. In this way he established the underwriting of thermodyamics by mechanics on the basis of the interplay between what we called (see above, Section 5.1) thermodynamic distinguishability and thermodynamic regularity.

5.7 Maxwell–Boltzmann distribution

In this section we show how the *Maxwell–Boltzmann* (MB) distribution of energy among the molecules of an ideal gas is derived in Boltzmann's combinatorial approach. In our approach, the MB distribution is a macrostate: all the microstates in this macrostate share a property, which is the MB energy distribution. We then explain in what sense this macro-state can be understood as the equilibrium macrostate. Unlike the stand-ard literature on this subject, in the following construction we avoid using the notion of probability. In particular we do not characterize the MB distribution as the most probable one. We shall rely only on combinator-ial reasoning. As we shall see in the next chapters, the notion of probabil-ity of a macrostate in statistical mechanics is independent of combinatorial considerations.

The MB distribution describes an ideal gas consisting of N molecules and total energy E. The number of molecules n_i in the gas that have their energy in the interval $[E_i, E_i + \Delta E]$ is called the *occupation number* of the energy interval $[E_i, E_i + \Delta E]$ (here the square brackets denote a closed interval). The gas is subject to the two following constraints. The total number of molecules is conserved:

$$N = \sum_{i=1}^{M} n_i,$$

and the total energy of the gas is also conserved, that is, the system is isolated:

$$E = \sum_{i=1}^{M} n_i E_i.$$

In thermodynamics, equilibrium is characterized as a state in which the magnitudes are constant over time. The derivation of the MB distribution begins with the reasonable assumption (or definition) according to which in a state of thermodynamic equilibrium the occupation numbers are constant. That is, the identities of the particles that have their energy in a given interval may change, but the number of molecules that goes into this interval must always be equal to the number that leaves it, so that the total occupation number of molecules is conserved.

Every microstate of the gas in the state space has some fixed energy. Therefore, a given energy interval $[E_i, E_i + \Delta E]$ corresponds to a continuum of microstates of the gas. In the MB construction the microstates in each energy interval are taken to be indistinguishable. Moreover, permutations of molecules within a given energy interval are likewise indistinguishable and taken to denote the same state. This means that the set of occupation numbers $\{n_i\}$ corresponds to a single arrangement of the gas. By contrast, permutations of molecules across different energy levels are also indistinguishable but counted as denoting different states. All the permutations of molecules within a given set $\{n_i\}$ therefore correspond to a macrostate of the gas – a distribution in Boltzmann's terms.

We now ask what the arrangement $\{n_i\}$ is of occupation numbers among the various different energy intervals, when the system is in equilibrium. Different arrangements are associated with a different number of permutations: there is, for example, an arrangement in which all the particles are within some given energy interval, and in this arrangement the number of permutations between energy intervals is zero; whereas all other arrangements, in which particles occupy different energy levels, are compatible with a positive number of permutations.

It turns out that the MB energy distribution has two distinct features. (i) Maxwell and Boltzmann derived successful empirical predictions from the MB energy distribution. (ii) The MB distribution is associated with the arrangement that is compatible with the *maximal* number of permutations. If we measure the size of macrostates by the number of permutations, the MB energy distribution will be the largest macrostate. It is important to distinguish between these two features of the MB distribution, since as we will show they are conceptually independent. Whether or not an equilibrium macrostate is invariably the largest macrostate (that is, whether or not equilibrium invariably has the maximal entropy) is a question we discuss later. Note that the question also depends on how size in the state space is measured.

Now, for each arrangement (i.e. energy occupation numbers) $\{n_i\}_j$ there corresponds a number of possible permutations between molecules belonging to *different* intervals. This number is given by:

$$N_{\text{per}}(\{n_i\}_j) = \left(\frac{N!}{n_1!n_2!\ldots n_M!}\right)_j.$$

(The subscript j indicates that the numbers n_1, n_2 etc. are of arrangement j.) Which arrangement corresponds to the maximal number of possible permutations? The total number of permutations in all arrangements is given by:

$$\sum_j N_{\text{per}}(\{n_i\}_j).$$

For a given system, this number is a constant, so the relative number of permutations belonging to an arrangement number j is:

$$P_j = \left(\frac{N!}{n_1!n_2!\ldots n_M!}\right)_j \times \frac{1}{\sum_j N_{\text{per}}(\{n_i\}_j)}.$$

Hereafter we omit the subscript j, for simplicity of notation. For the sake of convenience we now turn to examine $\log(P)$ instead of P. The reason is that the extrema of these expressions coincide, and it is more convenient to calculate the extremum of $\log P$. This immediately implies that

$$\log P = \log(N!) - \sum_{i=1}^{M} \log(n_i!) + \text{const.}$$

By Stirling's approximation,

$$n! \approx \sqrt{2\pi n}\left(\frac{n}{e}\right)^n$$

and so

$$\log(n!) \approx \left(n + \frac{1}{2}\right)\log n - n + \text{const.}$$

Substituting in the above expression for log P we get:

$$\log P \approx N \log N - \sum_{i=1}^{M} n_i \log n_i + \text{const.}$$

We now want to find the arrangement (or set of occupation numbers) $\{n_i\}$ for which this expression has its extremum. Using the method of Lagrange multipliers, we look for the extremum of log P under the constraints $N = \sum_{i=1}^{M} n_i$ and $E = \sum_{i=1}^{M} n_i E_i$. This yields:

$$\delta(\log P) = 0, \delta N = \sum \delta n_i = 0, \delta E = \sum E_i \delta n_i = 0$$

or

$$\sum_{i=1}^{M} (\log n_i + \alpha + \beta E_i) \delta n_i = 0,$$

where the δn_i are independent. We solve for each i:

$$\log n_i + \alpha + \beta E_i = 0$$

and get:

$$n_i = e^{-\alpha - \beta E_i}.$$

This set of occupation numbers $\{n_i\}$ gives the maximum of P. It is the Maxwell–Boltzmann distribution.

To interpret β, we first calculate an expression for the pressure of an ideal gas (pressure = exchange of momentum with walls), and get

$$p = \frac{N}{V\beta}.$$

Combining this expression with the thermodynamic equation of state for an ideal gas, $PV = NkT$ (where N is the number of molecules in the gas), we get

$$\beta = \frac{1}{kT}.$$

To obtain α, we use normalization considerations on the basis of the constant number of particles in the gas. Therefore, the relative number of molecules in each energy interval $[E_i, E_i + \Delta E]$ when the gas is in the macrostate called the MB energy distribution is:

$$\frac{n_i}{N} = Ce - E_i/kT(\text{where } C \text{ is a constant}).$$

From this we can calculate magnitudes pertaining to ideal gases, and compare them to the thermodynamic predictions.

5.8 The observer in statistical mechanics

Since in the discussion of macrostates the notion of an observer has a central role, and since this notion and its meaning in a discussion that aims to be purely physical may appear to create difficulties, we now turn to address several aspects of this notion, relevant to its role and status in statistical mechanics.

Classical mechanics is *a theory without an observer* that describes the evolution of microstates. It is without an observer in the sense that notions such as observation, measurement, and so on, have no primitive status in the theory, and are reducible (in principle) to the mechanical principles that govern and explain any other phenomenon, object, or process. Nevertheless, the notion of an observer appears in statistical mechanics in the following contexts. The first context is the very idea that science is empirical in some sense, which implicitly involves the notion of observation. This idea is not special to statistical mechanics or to mechanics itself: it is at the heart of the idea of an empirical science. Although this seems to be a somewhat abstract idea, we will see in Chapter 10 that it plays a significant role in accounting for the time-directedness of the thermodynamic laws.[17]

The second place where observers appear in mechanics is in forming the bridge between the atomistic hypothesis and the phenomena, where the notion of macrostates comes into play. Here, the important aspect of observers is that they are physically correlated with their environment in a one-to-many (or many-to-many) way, which brings about limited observation capabilities. In order for mechanics as well as statistical mechanics to remain theories without an observer, the observer that gives rise to macrostates has to be understood as a system that can be fully described in purely mechanical terms. This idea is called (in philosophy of mind)

---

[17] This idea underlies Albert's (2000, Ch. 4) so-called *Past Hypothesis*; see Chapter 10.

physicalism. We now turn to explain in outline how the idea of physicalism constrains the account of macrostates in mechanics.

Physicalism in philosophy of mind is the idea that the mental is physical, and nothing but physical.[18] Since our account of macrostates brings in the notion of an observer, the question arises as to whether this account is indeed part of statistical mechanics, in other words whether the notion of an observer is mechanistic. This brings in the idea of physicalism. As we emphasized, our notion of an observer does not imply anything non-physical about the properties of O in the sense that the properties of both O and G are only those that can be described in terms of their mechanical degrees of freedom. O and G have the same kind of properties, all of which can be described in purely mechanical terms, and both satisfy the same laws of physics. Nevertheless, the distinction between O (as an observer) and G is interesting precisely because O has an experience, and it is this fact that physicalism aims to explain.

Of course, the usual understanding is that not every microstate of every system is a *mental* state; only certain microstates of special kinds of systems, that have the appropriate structure, are mental states; human beings, in particular, have such a structure; and we may assume that O has such a structure. This means that in order for a physical system to be an observer, all it needs is the right physical structure; nothing extra-physical or super-physical is needed. In particular, the structure of O needs to be characterized, as a matter of principle, in purely mechanical terms. For if the terms describing the observer are not purely mechanical, then we cannot say that the underwriting of thermodynamics is done by mechanics, rather than by some other (possibly non-physical) theory, and in this case our account will not be physicalist.

To fix the ideas, consider Figures 5.2 and 5.3. Since according to physicalism mental states are nothing but physical states, we may say that when O is in the microstate x_1, O is in the *mental* state x_1. This physical state can *also* be described using terms belonging to the mental vocabulary such as "experience" without in any way undermining the physicalist idea. In this case we will say that O experiences that G is in the macrostate $[M_1]$.

---

[18] There are various ways of accounting for the relation between mental states and physical states. Here we phrase the discussion in terms of the so-called *identity* theory (type or token).

It is a well-known idea that mental states are multi-realizable by physical states. That is, one normally takes it that small changes in microscopic brain states, such as a movement of one atom in a molecule in the brain, do not bring about changes in the experienced mental states. Similarly, different people (and other animals) have different brain states, but we normally accept that they have similar (or at least comparable) mental states. And so it seems that the general picture describing the correlations between O and G should be the one illustrated by Figure 5.3. In this figure, each macrostate of G is associated with several, and possibly a continuum of, microstates of O. Whenever O is in *any* of the microstates in region O_1 (on the O axis), O is in the mental state of experiencing G in the macrostate $[M_1]$; and similarly for the regions O_2 and $[M_2]$. We call this state of affairs *multi-realizability*, since it expresses the idea that O's experience of a given macrostate of G can be realized by *different* microstates of O.

However, in order for multi-realizabilty to be embedded in physicalism, the region O_1 must be characterized in mechanical terms. That is, we need to assume that there is some mechanical property that is shared by all and only the microstates in O_1. Otherwise, if there is no such property, in particular if the only way to characterize the region O_1 would be extensionally, or alternatively by the function (or causal role) played by the microstates in O_1, then it is obvious that such a theory would not be fully mechanistic or even physicalist.[19]

It is a widely accepted view in statistical mechanics that identity statements of the form "heat is average kinetic energy" are true of our world. This view is taken over to philosophy of mind, for example by Saul Kripke.[20] Although Kripke is right in rejecting the idea that the reduction of thermodynamics to mechanics is analogous in some sense to the reduction of the mental to the physical, his reasons for this rejection are mistaken.

There are three different concepts involved in this discussion: molecular motion of the gas, the mechanical brain states of the observer, and the mental states of the observer (for example the sensation of

---

[19] The idea of multi-realizability came up explicitly for the first time in the context of physicalism by Hilary Putnam (see Putnam 1967). Putnam's paper is considered by many to be the central one in refuting the so-called identity theory in philosophy of mind and in leading the way to its replacement by some version of functionalism. We argue above that in the context of physicalism, functionalism, conversely, must eventually be reduced to some sort of an identity theory.

[20] See Kripke (1980, Lecture III).

heat). The identity statement "heat = molecular motion" is simply mistaken since, obviously, states of different objects cannot be identical[21]. As we said earlier, the molecular state of the gas and the mechanical brain states of the observer may be correlated (where the correlation may come about causally). In terms of Figure 5.3, each of the microstates in $[M_1]$ of G has the same average kinetic energy, and it is this physical feature or aspect of each of the microstates in $[M_1]$ which brings about (perhaps causally) one of the microstates in O_1 (say, O's brain state x_1). The idea that the reduction of thermodynamics to mechanics should be partly based on some conceptual analysis seems to be widely accepted in the philosophical literature. For the reasons spelled out above, this idea is wrong.

So what is heat? It is not a kind of motion[22] but rather a kind of *sensation* or a mental state, which is experienced by an observer.[23] Identity can only be between the microstate x_1 of O's brain and O's mental experience of heat. This is what one means in the context of the identity theory of mind when one says that x_1 is identical with O's mental state of experiencing heat.[24] However, in the philosophical literature the idea is that the identity or supervenience relation between mental states and physical (brain) states need be justified by *a priori* considerations. The idea, for example, is that the sensation of heat (or pain) can be said to supervene on some (kind of) brain states only if one can come up with some sort of conceptual analysis that leads to this identification. One of the main arguments supporting this idea is that in, for example, statistical mechanics it turns out that there is a conceptual linkage between heat and molecular motion. However, as we argued earlier, this is wrong (see the beginning of Section 5.5). As we saw, there is no *a priori* reason why a particular set of microstates of G should be correlated with a particular set of microstates of O rather than with any other. And therefore, *ipso facto*, there is no *a priori* reason why a

---

[21] Kripke understands the identity "heat=molecular motion" differently, i.e. as implying that heat is the name or the label for molecular motion. But in this case, the label "heat" is a place holder for whatever turns out to bring about the sensation of heat in us. If it turns out by empirical research that molecular motion of some sort plays this role, then the identity "heat=molecular motion" is true – but unlike Kripke, we say that it is true by convention (or definition).

[22] This is of course an allusion to Clausius's expression: "the kind of motion we call heat" (see Brush 1976).

[23] Compare Kripke (1980, Lecture III).

[24] In the terms used in philosophy of mind, Figure 5.2 may be associated with *token* identity theory, and Figure 5.3 with *type* identity.

particular set of microstates of G should be correlated with some (kind of) mental state (or experience) of O.[25]

We believe that a physicalist account of the observer is necessary for a complete underwriting of thermodynamics by mechanics. Of course there is no reason to hope that the physicalist project can be carried out in full by classical mechanics since classical mechanics is not the fundamental physical theory of the world. At the same time, even present-day physics is very far from a complete account of the mental. But the very idea of physicalism is in general to account for mental states and the experience of observers by a physical theory without an observer. In statistical mechanics the observer is essential but, nonetheless, assuming a physicalist account of the observer, the theory can be made fundamentally *observerless*.

5.9 Counterfactual observers

Sometimes it seems *as if* statistical mechanics can be made observerless, at least in those cases where thermodynamic properties and macrostates are assigned to systems in scenarios in which observers are absent. But this appearance is misleading: if indeed no observer is involved, then mechanics tells us that the only right way to describe the state of a system would be in terms of its actual *micro*state. Assigning macrostates to a system in cases where observers are absent can mean only postulating a *counterfactual observer*, and asking what that counterfactual observer would experience, were it present in the scenario, and had it observed the system of interest. Of course, we can assign to such an imaginary observer any observation capabilities we wish, and thus partition the state space into macrostates in any way we like. The fact that we normally choose a partitioning in thermodynamic macrostates attests that we normally have in mind an imaginary human observer, aided by some known measuring devices, and we use macrostates that fit such an observer. This is the meaning of assigning, for example, thermodynamic magnitudes to objects which are far away or long gone. Of course the structure of trajectories in the state space of an observerless scenario need not be different from the structure of the trajectories of observed systems. But without an observer,

---

[25] Perhaps it will turn out as a matter of contingent fact that our sensation of heat is identical to (or supervenes on) some aspects of the motion of the particles that make up our brains. But this is altogether a different story.

either actual or counterfactual, it is meaningless to ascribe macroscopic regularities to mechanical systems.

One implication of the fact that an assignment of macrostates necessarily assumes an observer, at least counterfactually, is that the notion of macrostates does not apply to the universe as a whole: if the whole universe is our observed system G, then no degrees of freedom are left for our physical observer O, and hence the assignment of macrostates is meaningless from a physicalist point of view; to the universe as a whole only *micro*states can be assigned.[26]

---

[26] This does not really pose a threat to the project of underwriting thermodynamics by statistical mechanics, since in *thermodynamics*, temperature – and *ipso facto* entropy – is undefined for the entire universe. (Our argument here is different from the argument by Ridderbos and Redhead (1998) for interventionism in the Gibbsian approach.)

6

Probability

6.1 Introduction

We have, up to now, constructed two mechanical notions over and above the elementary mechanical ones. These are the notions of a macrostate and of a dynamical blob. We also saw that these two notions are not sufficient to underwrite the time-directed laws of thermodynamics by mechanics. We conjectured that mechanical counterparts of the laws of thermodynamics would have to involve the notion of probability. In this chapter we construct this notion.

Historically, Boltzmann first attempted to prove a *deterministic* mechanical counterpart of the thermodynamic Law of Approach to Equilibrium: he tried to show that every thermodynamic system inevitably ends up in equilibrium, regardless of the details of its initial state or its evolution. However, Boltzmann had to abandon this deterministic project, following the counter-examples put forward by Loschmidt and Zermelo (see Chapters 3 and 7).[1]

Boltzmann's insight in the face of these objections was his so-called *probabilistic turn*, in which he attempted to show that although the laws of thermodynamics cannot be derived from mechanics, *probabilistic* counterparts of the laws of thermodynamics *are* derivable from mechanics. He tried to prove that anti-thermodynamic evolutions – that is, evolutions away from equilibrium – are not strictly impossible but extremely improbable, and therefore in practice we ought not to expect to see them. Hence, in practice, according to Boltzmann's argument, the case in which anti-thermodynamic evolutions are merely improbable is empirically indistinguishable from the case in which such evolutions are strictly

[1] See Uffinle (2006, See 4.3).

impossible. The central aim of statistical mechanics is to prove Boltzmann's intuition.

Introducing probability into statistical mechanics is subtle, especially because classical mechanics is a deterministic theory. There are two desiderata that we believe the notion of probability in statistical mechanics must satisfy. (i) Probability should be characterized in purely *mechanical* terms. In particular we construct the notion of probability on the basis of the interplay between macrostates and dynamical blobs as developed so far. A notion of probability based on mechanical terms is mandatory in order to describe Boltzmann's insight (let alone prove it) in mechanics. (ii) We shall take it here that probabilistic statements must be *empirically* testable to the same extent as non-probabilistic statements such as Newton's Second Law, $F = ma$, Maxwell's equation of the magnetic field, or the Schroedinger equation in quantum mechanics.[2] We call this the *testability criterion*. Assuming that probabilistic counterparts of the laws of thermodynamics that satisfy the above two desiderata can be formulated, we shall say that they are as physically objective as mechanics and thermodynamics are.

6.2 Probability in statistical mechanics

In a nutshell, probability enters statistical mechanics at the interface between macrostates and dynamical blobs owing to the fact that as the blob evolves according to the dynamics of the system, in general it partly overlaps with different macrostates. These partial overlaps between the blob and the macrostates determine the probabilities of future macrostates, and so in order to make probabilistic predictions one can follow the evolution of the blob by solving the mechanical equations of motion for the initial microstates contained in the initial macrostate, and measure the partial overlaps of this blob at the desirable future time. In this section we will describe in detail this idea and some of its consequences.

A very important remark

There is a very important remark we wish to state at the outset. Since the probability of future macrostates depends on the evolution of the blob, one might conclude that to make statistical mechanical predictions

---

[2] We shall not go into the epistemological and methodological issues that arise in the context of the problem of empirical testability of general theoretical statements.

one must actually carry out detailed calculations of that evolution. But such detailed calculations are not feasible, owing to the degree of precision and to the number of particles and the number of equations involved. Consequently, one might suspect that our account of probability (described in a nutshell above) cannot serve as the basis for understanding statistical mechanics. But this would be a mistake: although the concept of probability in statistical mechanics is indeed based on dynamical blobs, the actual dynamics of our universe is such that shortcuts are possible, in which one can come up with probabilities without the need to carry out detailed blob calculations. We discuss these shortcuts in Section 6.5 below.

But whether or not there are such shortcuts, we will show that probability in classical mechanics entirely *supervenes* on the interplay between blobs and macrostates. Moreover, the very idea of attempting to prove in mechanics probabilistic counterparts of the laws of thermodynamics requires that our notion of probability be grounded in the mechanical terms of macrostates and dynamical blobs. From a theoretical point of view, there are no conceptual shortcuts. And so, we will first show how probabilities arise in statistical mechanics through the relation between dynamical blobs and macrostates, and then continue to see what exactly is the nature of the shortcuts which – luckily – save us the trouble of going through cumbersome calculations. (Since this remark is extremely important, we shall refer to it several times as we proceed as "the very important remark in Section 6.2.")

To illustrate the discussion, consider the paradigmatic example of a gas in a container, which is initially, at time t_0, confined by a partition to the left-hand side of the volume of a container; see Figure 1.1. At time t_1 the partition is removed and the gas is allowed to evolve spontaneously. We know from experience that the gas will quickly expand, and after a short time interval it will fill the entire volume of the container, as shown in the figure.

Here is the state space description of this scenario. Figure 6.1 describes the state of the world at t_0, when the gas is still confined to the leftmost quarter of the container by the partition. In the figure, for obvious reasons, we can depict only three axes, and therefore we represent the gas G by two degrees of freedom G_1 and G_2, and we assume that they are representative of the general qualitative behavior of the other degrees of freedom of the gas; and O is, as usual, the observer, relative to which the macrostates of the gas are defined. The accessible region of $O + G$ resides in their combined multi-dimensional

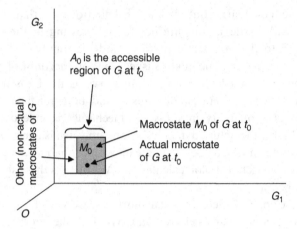

Figure 6.1 State-space description of a gas confined to the left-hand side of a container

state space; but here we shall be interested only in the accessible region of the gas G. We shall, therefore, focus on the projection of the accessible region on the G subspace: in Figure 6.1, the accessible region of the gas G is the region A_0 in the $G_1 + G_2$ plane. By assumption, the macrostate of G (relative to O) at t_0 is $[M_0]$, in which the gas fills the entire leftmost quarter of the container (recall that we denote macrostate regions by square brackets). The part of the accessible region A_0 which is outside $[M_0]$ contains other macrostates of the gas, such as that in which the gas fills only $1/8$ of the volume of the container; but in our story it happens to be the case that as a matter of fact the microstate of the gas at t_0 is within macrostate $[M_0]$. Assuming that the gas satisfies the thermodynamic Law of Approach to Equilibrium, and that it has been confined by the partition to one-quarter of the container for some time, we may assume that $[M_0]$ is the equilibrium macrostate relative to the constraint of this partition.

At t_1 the partition is removed. This change of constraints means that at t_1 the accessible region of the gas changes *instantly* from A_0 to the larger region A_1, which includes the region A_0; see Figure 6.2. The new accessible region A_1 includes $[M_0]$ as well as other macrostates, such as $[M_1]$ and $[M_2]$ in Figure 6.2. For example, $[M_1]$ could be the macrostate corresponding to the gas filling the left three-quarters of the volume of the container, and $[M_2]$ could be the macrostate corresponding to the gas filling the entire volume of the container which, according to our experience, is the equilibrium macrostate of such systems.

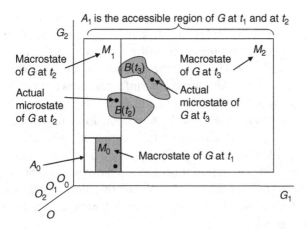

Figure 6.2 State-space description of a gas expanding in a container

The sets of microstates $[M_0]$, $[M_1]$, and $[M_2]$ are macrostates of G because the observer O can distinguish between these three sets, but cannot distinguish between the microstates within each set: O can tell whether the actual microstate of G is within $[M_0]$ or $[M_1]$ or $[M_2]$, but cannot tell which is the actual microstate within each of them. As we explained in Chapter 5, this fact is determined by the structure of the accessible region of $O + G$ in the composite state space, which brings about the right sort of correlations between the microstates of O (denoted in the figure by O_0, O_1, and O_2) and the microstates of G that are within $[M_0]$, $[M_1]$, and $[M_2]$, respectively. Of course, the structure of the accessible region, and *ipso facto* the partition to macrostates, is determined by the physical constraints on O and G and the physical interactions between them.

Immediately after the removal of the partition, at t_1, the actual microstate of G is still somewhere within the region $[M_0]$, since the gas is still in the leftmost quarter of the container. Now suppose that at that time, namely at t_1, we wish to *predict* the macrostate that the gas will have at some later time, t_2.

(Terminological remark: In making the prediction, we are going to take the point of view of the observer O in the sense that we shall take the information available to us about the gas at each time to be the information available to O. To this information all we need to add is our knowledge of the equations of motion, on the basis of which we can calculate the evolution of the dynamical blob. Thus whenever we say that *we* make a prediction, we mean that it is O who is making the predictions that we describe. For ease of presentation only, we will say *we*.)

If we knew the actual microstate of G at t_1, we could in principle (complexity notwithstanding) calculate the trajectory that starts out at that microstate, and find the exact future microstate of G at t_2, and thus predict its macrostate. However, the assumption from which statistical mechanics starts is that this microscopic information is not available to us, and all we know is that the microstate of G at t_1 is somewhere within the macrostate $[M_0]$. Therefore, the best we can do is to calculate *all* the trajectory segments that start in *all* the microstates within $[M_0]$, and see where this bundle of trajectories will end at t_2. The region that includes all the end points of this bundle of trajectories is the dynamical blob of G at t_2, given that G started out at t_1 in $[M_0]$. We call this blob $B(t_2 | [M_0], t_1)$ or, for simplicity of notation, $B(t_2)$, where we omit mentioning t_1 and $[M_0]$ but keep them in mind.

By construction, the dynamical blob at the beginning of the evolution is always the initial macrostate. Here, $B(t_1) = [M_0]$, that is, the dynamical blob at the initial time t_1 (right after the removal of the partition) coincides with the initial macrostate at that time, $[M_0]$. By contrast, when we turn to calculate the time-evolved blob $B(t_2)$, we find that in general it partly overlaps with several macrostates. In the case described in Figure 6.2, the blob $B(t_2)$ partly overlaps with $[M_1]$ and $[M_2]$. Now, since the gas G could have started in any of the microstates in $[M_0]$, it can end in any of the points in $B(t_2)$ – either in a point that is within $[M_1]$ or in a point that is within $[M_2]$. Since we do not know which was the actual microstate within $[M_0]$, this prediction does not tell us which of the two macrostates $[M_1]$ or $[M_2]$ will be the actual case. And it is here that probability enters the picture; it is here that we begin to deal with *statistical* mechanics proper.

Since the maximal information that O has about G at t_0 is $[M_0]$, all that O can calculate about G's state at t_2 is expressed by $B(t_2)$, and therefore the best that O can say about G's future state at t_2 is in terms of the probability that G will end up in any of the macrostates $[M_1]$ or $[M_2]$. In other words, O can – at best – assign probability to G ending up in either $[M_1]$ or $[M_2]$ at t_2, given that G started out in $[M_0]$ at t_1. Let us see how these probabilities can be calculated, and what they mean.

The probability that at t_2 the macrostate of G will be $[M_1]$, namely that the actual microstate of G at t_2 will be within the region $[M_1]$, given that at t_1 it was in $[M_0]$, is called the *transition probability* from $[M_0]$ at t_1 to $[M_1]$ at t_2; and this transition probability is given by the *relative size of the overlap* between the dynamical blob $B(t_2)$ and the macrostate $[M_1]$ (see Figure 6.2).

Formally, the rule for calculating transition probabilities in statistical mechanics is the following:

(Probability Rule) $P([M_1], t_2|[M_0], t_1) = \mu(B(t_2) \cap [M_1])$.

This means that the probability that a system that starts in macrostate $[M_0]$ at time t_1 will end in macrostate $[M_1]$ at time t_2 is given by the relative size μ of the dynamical blob $B(t_2)$ which overlaps with the macrostate $[M_1]$. (Of course, in order to have probability here, the relative size μ of the overlap has to be normalized, so that the sum of the overlaps of the blob $B(t_2)$ with all the macrostates at t_2 will be equal to 1. From now on we will assume that the μ measure is normalized. More details concerning the μ function are discussed in the next section.)

This is how probability arises in a deterministic world. The transition probabilities are an objective feature of the physical world, despite the determinism of the Newtonian trajectories, since probabilities express the relations between macrostates and dynamical blobs, both of which are objective features of the world. Macrostates are objective since they are the outcome of the physical correlations between the observer O and the observed system G expressed in the structure of the accessible region of $O + G$, and are determined by the physical constraints on and the physical interaction between O and G. And dynamical blobs are objective since they are fixed by the equations of motion, given the initial macrostate. Consequently, whether or not a blob evolves such that at some time it is spread over more than one macrostate is an objective feature of the world. Whenever the evolution of the blob turns out, as a matter of fact, to split over more than one macrostate, probability objectively comes into play. Since both the notion of a blob and the notion of a macrostate express ignorance (see Section 5.1), the probabilities in statistical mechanics express *objective ignorance*.

Before we continue further with details concerning the concept of probability in statistical mechanics as based on the Probability Rule, it is important to recall the very important remark we made at the beginning of this section, namely that this concept is indeed the way in which probability arises in statistical mechanics, although as a matter of fact we are unable to carry out the calculation of probabilities according to the Probability Rule. This is because we do not have the computational capability that is required in order to calculate the evolution of the dynamical blob, and often we also lack information on the details of the partitioning of the state space to macrostates. In Section 6.8 below we address this point and explain how probability calculations ought to be carried out in *practice*, based on the Probability Rule and taking into account the above limitations. But, before doing so, there are some points concerning the above notion of probability that need clarification.

6.3 Choice of measure in statistical mechanics

The formal expression of the Probability Rule is not complete unless we specify the function μ, and this point turns out to be highly non-trivial. Normally in statistical mechanics μ is taken to be the Lebesgue measure, which is closest to the intuitive way in which we usually think about the notion of size: for example, the Lebesgue measure of the interval $[a,b]$ is $b - a$. However, the mathematical theory of measure abstracts from the intuitive notion of size a notion of measure, by specifying the conditions in which a function is a measure;[3] and the Lebesgue measure is just one of them. There are other functions, satisfying the mathematical criteria for being a measure, that yield different values for the size of $[a,b]$. An example of a measure function is the Dirac delta measure, according to which the whole measure is, as it were, concentrated in one single point, and the size of an interval $[a,b]$ is 1 or 0 depending on whether or not that point belongs to that interval.[4]

Which measure should be used as μ in the Probability Rule? To answer this question we apply the guideline presented in the introduction to this chapter, which we called the *testability criterion*. In physics, the choice of measure for calculating probabilities should have *empirical significance*, in the sense that the adequacy of this choice ought to be empirically testable, no less than any other statement of physics. With this guideline in mind, we propose the following procedure for choosing the measure

---

[3] A measure is a function defined over a collection of sets which forms a σ-alebera, that is, a non-empty collection of subsets of a given set X, closed under complementation, union, and intersection. In particular, the collection includes X itself and its complement, the empty set $\emptyset$. In this case, the function μ from X to the Reals is a measure if it satisfies the following three axioms:

   (I) Non-negativity: for all subsets x_i of X, $\mu(x_i) \geq 0$.
   (II) Countable additivity: μ of a union of sets is equal to the union of μ of the sets: $\mu(\cap x_i) = \cap(\mu(x_i))$.
   (III) $\mu(\emptyset) = 0$. A measure is a probability measure if it is from X to the interval $[0,1]$ of the Reals, and then (III) entails $\mu(X) = 1$.

[4] Another example due to Albert (2000, Ch. 3) is this. The magnitude $b^2 - a^2$ satisfies the conditions on a measure function, but yields different values for the distance between a and b: the measure of $[\frac{1}{2},1]$, for instance, is $1 - \frac{1}{2} = \frac{1}{2}$ by the first measure, and $1 - \frac{1}{4} = \frac{3}{4}$ by the second measure.

μ to be used in the Probability Rule; we present this procedure in terms of the example of Figures 6.1 and 6.2 for clarity and concreteness, but the procedure is quite general.

Suppose that we prepare a large collection of systems that are identical to the gas G in their thermodynamic parameters and in the thermodynamic constraints that act on them, and all of them start out in the macrostate in which the gas fills the leftmost quarter of the container.[5] Formally, all these systems are represented in the same state space, all of them have the same accessible region, and all are initially in the same thermodynamic macrostate $[M_0]$. After removing the partition we let these gases evolve from time t_1 to time t_2. We notice the macrostate to which each gas arrives at t_2, and count the observed relative frequencies of these macrostates among all the gases. Assuming that our calculations, expressed by the dynamical blob $B(t_2)$ in Figure 6.2, were correct, the final macrostate of some of the gases will be $[M_1]$ and the final macrostate of the other gases will be $[M_2]$, and we denote their observed relative frequencies by F_1 and F_2, respectively. We now choose a measure μ, according to which the relative measures of the overlap between the calculated dynamical blob $B(t_2)$ and the macrostates $[M_1]$ and $[M_2]$ will be, to an acceptable approximation, equal to the observed relative frequencies F_1 and F_2, respectively.[6] (Alternatively, we start with some prior μ, and the relative frequencies F_1 and F_2 ought to be the basis for updating μ. Here we shall not make the (conceptually important) distinction between these alternatives: whenever we say that μ ought to be chosen so as to fit the observed relative frequencies, we shall mean that this can be done in *any* of the ways for adapting probabilities to observed relative frequencies.) In general, there may be many measures that satisfy this condition. Among those measures we choose one that is most *convenient*; for instance, it may be simple, in some sense of simplicity (see below).

This way of choosing μ is empirically testable since if, in future experiments, we find relative frequencies that significantly differ from the probabilities based on the chosen μ, we should change the measure μ in the Probability Rule, so that the new measure will reflect the updated observed data. The term "significantly" here is to be understood as in the

---

[5] Here we treat the gases as a *collection* of systems rather than as an ensemble. See Peres (1993, p. 25 n.1; and p. 59 n.9) for the distinction between a collection (or an assembly) and an ensemble.

[6] Compare the section "The Relation to Experimental Data" in Kolmogorov (1933, pp. 3–5).

usual statistical methods, and the update of the probabilistic statement embodied by μ is to be carried out in the usual ways. In this way we obtain a measure μ that fits our experience and is empirically testable and revisable, and hence is empirically significant.

This sort of considerations does not yield a unique choice of measure. This non-uniqueness is part of the well-known and unavoidable under-determination of theory choice by empirical data in physics. However, our procedure for choosing a measure fares, in this respect, as well as the choice of non-probabilistic laws of nature such as $F = ma$. It does not fare worse than $F = ma$ since both are underdetermined by the observed data; and it does not fare better than $F = ma$ since both are based on experience. (In Chapter 8 we shall address approaches according to which probability assignments can fare better than $F = ma$.)

As we said above, an explicit part of the process of choosing the measure μ in the Probability Rule is to choose the most convenient measure among those that fit the observed relative frequencies. Conveni-ence is perhaps a vague and subjective notion, so that different people may prefer different measures; but as long as the chosen measure fits the observed relative frequencies, this non-uniqueness is insignificant.

Still, there is one convenience consideration that is very often used, albeit not explicitly, and is worth mentioning: it is the requirement that relative to the chosen measure, the weights of the different regions of the blobs at all times be distributed uniformly. It is important to realize that the choice dictated by the relative frequencies is actually a choice of *a pair of measures and a distribution of weight*; and when we talk about a choice of a measure only, we normally intend, implicitly, that the distribution of weight will be the most convenient one relative to this measure, namely, uniform. A uniform distribution is one that gives equal weights to regions that are equal by a given measure. Although choosing a measure so as to have uniform distribution of weights relative to it is certainly convenient, it is not necessary, and nothing in mechanics gives it any *a priori* prefer-ence.[7] Any convenient pair of measure and distribution that fits the observed relative frequencies is acceptable.

Nevertheless, the Probability Rule mentions only the measure, and so it already contains the assumption that we are going to choose a measure relative to which the weight will be distributed uniformly. Since this assumption is not necessary, one may replace the Probability Rule with a more complex one, one that will allow for different distributions. Here,

---

[7] In particular, we see no metaphysical advantage to any Principle of Indifference.

for simplicity, we use the uniform distribution assumption and so remain with the Probability Rule as presented above.[8]

The above procedure for choosing the function μ in the Probability Rule emphasizes the difference between probability and measure. Regardless of the way in which one interprets the concept of probability, one thing is clear: if probability has anything to do with empirical facts – and in physics probability should be *closely* connected with empirical facts – then probability ought to reflect[9] the observed finite relative frequencies of events. The Probability Rule yields the probability P only if we substitute for μ the right sort of function, one that reflects the observations. If it does not reflect the observations, we shall not call it probability in the context of physics, although it may satisfy the axioms of probability theory.

These guidelines to choosing the measure of probability are different from some of the mainstream ideas in the literature on this subject. One can roughly distinguish between three kinds of criteria to choose the measure of probability.

(i) Some of the criteria found in the literature are of an *a priori* nature. An example is Pitowsky (2012), who prefers the Lebesgue measure since he takes it to be a generalization of the counting measure, which he takes to be natural. (ii) Other proposed criteria appear to be dynamical. One such approach prefers the Lebesgue measure owing to the fact that it is conserved under the dynamics in classical mechanics. This consideration, however, is not relevant for the choice of μ in the Probability Rule. We expand on this point in Chapter 8. (iii) The third kind of criteria are empirical and pragmatic, and here we classify our own approach: we select a set of measures that yield predictions of finite relative frequencies which reproduce the past observed relative frequencies. For the choice of one measure among them, pragmatic criteria are applied. Allegedly, in all three approaches, the choice of measure is tested against experience. However, as we will show in Chapters 7 and 8, some of these choices have no empirical basis, and sometimes they even go against the empirical observations. This statement may sound surprising, and to understand its

---

[8] Note that this Probability Rule does not assume the Principle of Indifference since it assumes uniform distribution only for the sake of convenience.

[9] Probability reflects the observed relative frequencies, but is not (necessarily) identical to the observed finite relative frequencies, owing to the effect of the priors and to the process of updating. This non-identity is also a consequence of the fact that probabilistic statements are consistent with the occurrence of states of affairs that have low probability.

full meaning one has to go slowly through the construction of statistical mechanics. This will be described in the next few chapters.

Our approach to choosing a measure involves the conjecture that future relative frequencies will resemble past relative frequencies of macrostates, and in this sense it involves some sort of inductive infer-ence.[10] This poses no problem, since the justification of this process is not weaker than the justification of the way in which we come up with any other *non*-probabilistic theories of physics, such as the equations of motion in classical (or quantum) mechanics.

To determine the measure of probability μ in the Probability Rule, we observed a large collection of identical systems all of which were prepared in the same initial macrostate. We now turn to examine the way in which the Probability Rule with the chosen μ can be used to predict the evolu-tion of individual systems.

6.4 Measure of a macrostate and its probability

Suppose that we choose the right sort of measure μ, so that it yields successful predictions of future relative frequencies of macrostates. Then it immediately follows from the Probability Rule that the probabilities of macrostates are associated with the μ-measure of the *overlaps* between the blob and the different macrostates. In particular, the probabilities of macrostates need not in general be equal to the μ-measure of the macro-states themselves. Since it may happen that a relatively large part of the blob will overlap with a relatively small macrostate, there is no reason whatsoever – from what we have said so far – to suppose that a macro-state that has a large μ-measure will also be highly probable.

This idea that larger macrostates are invariably associated with a higher probability would entail the prediction that systems evolve directly to the macrostate with the largest measure. This prediction would be of course false for the following reason. By and large, the observed phenom-enon in thermodynamic systems is that systems evolve to equilibrium via macrostates that have gradually increasing sizes. Therefore, this equilib-rium ought to receive high probability. But the above prediction gives it

---

[10] Here we make no preference for inductive approaches over non-inductive ones. If appropriate, replace our mentioning of induction with whatever the preferred scientific methodology is. The main point is that the same methodology should be used in choosing μ as in choosing, say, the law $F = ma$.

only small probability.[11] Moreover, equating the probability of a macrostate with its size is also in clear contradiction to the laws of dynamics that govern the evolution of mechanical systems by dictating their trajectories. In fact, this idea may even entail a violation of the principles of special relativity, since it requires velocities higher than the speed of light (see also Section 7.10). By contrast, the Probability Rule takes into account the dynamics in a way that meshes very well with the fact that systems evolve to equilibrium gradually, that is via macrostates with increasing entropies.

The fact that the probability of a macrostate is generally independent of the size of that macrostate entails that the probability is also independent of the entropy of that macrostate. (We assume here, of course, that the statistical mechanical entropy of a system which is in a macrostate is given by the Lebesgue measure of that macrostate; see Chapter 7.) Indeed, it sometimes seems as though the fact that entropy is given by the Lebesgue measure is taken to be a reason for preferring the Lebesgue measure as the measure of probability. Despite the tempting convenience, however, the former use of this measure does not logically entail the latter.

Moreover, if indeed the Lebesgue measure were chosen as the measure of probability just because it is also the measure of entropy, then the statistical mechanical counterpart of the Law of Approach to Equilibrium and of the Second Law of Thermodynamics would be *true by definition*, rather than by contingent facts that call for explanation. In other words, the idea that large macrostates invariably have high probability entails that the central laws of thermodynamics are not generalizations from experience, but *a priori* non-physical truths.

6.5 Transition probabilities without blobs

At the beginning of Section 6.2 above, we made the following very important remark: despite the fact that the probability of future macrostates depends on the evolution of the blob, we often make probabilistic predictions concerning thermodynamic systems without going into detailed calculations of this evolution. Moreover, it would be quite unreasonable to demand that actual calculations of probabilities rely on blob

---

[11] We speak here of phenomena and not of the Law of Approach to Equilibrium since this law refers only to the final equilibrium state of the system, and not to the intermediate states through which the system passes on its way to equilibrium.

calculations since in general we do not have a full and precise description of the sets of microstates that make up the macrostates, and, moreover, the calculation of the evolution of the dynamical blob and its overlap with the different macrostates exceeds our computational capabilities. Since, as a matter of fact, the thermodynamic behavior in our experience is regular, we can infer the probabilities of macrostates from their relative frequencies. Consider the paradigmatic case of a gas expanding in a container. We prepare a large collection of such gases, all of them in the same macrostate $[M_0]$ at t_0. At some later time t_1 we count the relative frequencies of the different macrostates $[M_i]$ of the gas. These relative frequencies are the basis for calculating probabilities: if the relative frequencies are

$$\mathrm{RF}(M_i, t_2 | M_0, t_1) = a_i,$$

then the transition probabilities based on the finite relative frequencies will be

$$P(M_i, t_2 | M_0, t_1) \approx a_i.$$

This inference from the observed relative frequencies to the transition probabilities is inductive: it is a purely empirical generalization, based on the conjecture that past relative frequencies will be repeated in the future.[12] Since this generalization is expressed in terms of macrostates only and makes no reference to blobs or to the Probability Rule, a question arises: what is the role of the Probability Rule in statistical mechanics? The answer is that the Probability Rule provides a *dynamical* explanation in terms of the structure of trajectories in the state space for the success of generalizations from the observed relative frequencies to probabilities. In fact, without the Probability Rule we have no mechanical underwriting of thermodynamics.

Note that the Probability Rule itself is not complete unless we specify the measure μ. And the choice of μ is based on experience: we choose a measure μ so that the probabilities in the Probability Rule will match the relative frequencies in our experience. And so it turns out that the rule for calculating probabilities in statistical mechanics is itself grounded in both the underlying dynamics of the system, and in experience.

In terms of the Probability Rule we now understand the success of the thermodynamic generalizations as a result of a lucky harmony between

---

[12] This inference is as valid as any other non-probabilistic inference in science.

the underlying dynamics and the carving up of the state space into thermodynamic macrostates. We say "lucky" harmony since it is a fortunate matter of fact rather than any *a priori* truth about the world, or even a theorem that follows from the principles of mechanics.[13]

6.6 Dependence on observed history?

Our above account of probability in statistical mechanics implies that the probability assigned to future macrostates depends on the initial macrostate that one assigns to the system, and on the calculation of the dynamical blob, given that initial macrostate. In this section we will show that this means that the probabilities that one assigns to future macrostates may depend on one's observed history, and therefore may differ from one observer to another, even in the case where both observers correlate with the observed system in a way that brings about similar macrostates. In this section we explain this idea in detail. We will see in Section 6.7 using the spin echo experiments that the probabilities of macrostates do indeed depend on observed history, and then in Section 6.8 we will explain how in thermodynamics the probabilities of macrostates turn out to be independent of the observed history.

Consider again Figure 6.2. At time t_1 the information that the observer O has about the gas G is that G is in the macrostate $[M_0]$, and on the basis of this information O calculates that the dynamical blob at t_2 will be $B(t_2)$, so that at t_2 the gas will be in either the macrostate $[M_1]$ or the macrostate $[M_2]$, with probabilities determined by the Probability Rule, with some appropriate measure μ. In this scenario, for every macrostate $[M_i]$ we have:

$$P([M_i], t_2 \,|\, [M_0], t_1) = \mu(B(t_2) \cap M_i).$$

Similarly, if O wants to predict the macrostates that G can have at a later time t_3 and their probabilities (having, as before, only the information that at t_1 the macrostate of G was $[M_0]$), O will carry out the same algorithm. So O will calculate the evolution of $B(t_2)$ from time t_2 to time t_3, obtain the dynamical blob $B(t_3)$, and calculate the measure μ of the partial overlaps between $B(t_3)$ and the different macrostates, according to the Probability Rule: for every macrostate M_i we have:

---

[13] To the extent that evolutionary-type explanations have an explanatory value, we are happy to bring them in; see also Section 5.5.

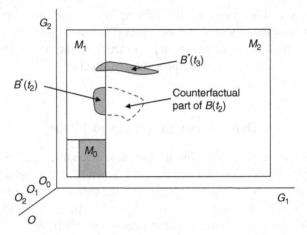

Figure 6.3 Collapse in classical measurement

$$P([M_i], t_3|[M_0], t_1) = P([M_i], t_3|B(t_2), t_2) = \mu(B(t_3) \cap M_i).$$

But suppose now that at time t_2 the observer O carries out a measurement of the actual macrostate of the gas, and discovers that the macrostate of G at t_2 is $[M_1]$, and not $[M_2]$. This means that the actual microstate of G at t_2 turns out to be within the region of overlap between $B(t_2)$ and $[M_1]$; call this region $B^*(t_2)$ (see Figure 6.3). O concludes that the region of overlap between $B(t_2)$ and $[M_2]$ is empty, in the sense that O now knows that it does not contain the actual microstate of G at t_2. The knowledge that the microstate of G is within $B^*(t_2)$ is a combination of three items: (i) the *observation* that at t_1 the macrostate of G was $[M_0]$, (ii) the *observation* that at t_2 the macrostate of G was $[M_1]$, and (iii) the *calculation* that, owing to the equation of motion of G, the dynamical blob that starts out in $[M_0]$ at t_1 evolves into $B(t_2)$ at t_2.

Because, before the observation at t_2, O's information about G's state was disjunctive, namely O knew that G was in either $B^*(t_2)$ or $B(t_2) \cap [M_2]$, and after that observation, O knows that G is in $B^*(t_2)$, we may say – cautiously borrowing a term from standard quantum mechanics – that measuring the macrostate of G at t_2 brings about an *epistemic collapse* of G's macrostate: the collapse is from the disjunction $B^*(t_2) \cup \{B(t_2) \cap [M_2]\}$ to the disjunct $B^*(t_2)$. In this collapse the microstate of G does not change,[14] but the state of O changes;[15] the collapse is perfectly *objective and physical* since it

---

[14] Unlike the collapse of the quantum state in standard quantum mechanics.
[15] We make here the idealization that the measurement does not alter the state of the measured system, as is usual to assume in classical mechanics.

reflects the one-to-many (or many-to-many) correlations between the micro-states of O and the microstates of G, brought about by their physical interactions. (We discuss in more detail this notion of collapse in Chapter 9.)

The collapse at t_2 has the following consequence. Suppose that at time t_2, after the epistemic collapse, O wishes to calculate the future evolution of G, in order to predict G's possible macrostates at some later time t_3 and their probabilities. Now, it would be pointless to calculate the evolution of the entire blob $B(t_2)$ since O already knows that the intersection of $B(t_2)$ and $[M_2]$ does not contain G's microstate. Therefore, in order to obtain the right prediction concerning the macrostates of G at time t_3, O needs to calculate only the evolution of the region that contains G's actual micro-state, namely $B^*(t_2)$. The region $B^*(t_2)$ is now the new basis for calculation of the dynamical evolution of G; from t_2 onwards, $B^*(t_2)$ replaces $B(t_2)$ in this role. The region $B^*(t_2)$ evolves to $B^*(t_3)$ (we denote their connection by marking both of them with asterisks), as illustrated in Figure 6.3, and the probabilities of the macrostates $[M_1]$ and $[M_2]$ are calculated as usual, using the Probability Rule. That is, we have for every macrostate $[M_i]$:

$$P([M_i], t_3 | B*(t_2), t_2) = \mu(B*(t_3) \cap M_i).$$

Obviously, if the outcome of the measurement at t_2 had been $[M_2]$, then the blob at t_3 would have been different, as would the probabilities for the macrostates at t_3.

And so we see that O assigns two different probability distributions over G's predicted macrostates at t_3, depending on O's observed history. In the first case, when O carries out a prediction of the macrostates of G at t_3 and their probabilities, relying only on the information that G's macrostate at t_1 was $[M_0]$, the probabilities at t_3 calculated by O depend on the blob $B(t_3)$. In the second case, when O carries out a prediction of the macrostates of G at t_3 and their probabilities, relying on the information that G's macrostate at t_2 was $[M_1]$, the probabilities at t_3 depend on the blob $B^*(t_3)$. In this sense, history matters in statistical mechanics or, more precisely, *observed history matters*.

Since, in general, different observers may differ in their observed histories, they may attach different probabilities to future occurrences. To see this, consider a different observer Q who is similar to O with respect to the correlations with G, such that O and Q assign the same macrostates to G for any microstate of G. Suppose, however, that Q joins our experiment only at t_2, and therefore does not know that at time t_1 the macrostate of G was $[M_0]$. Therefore Q is unable to calculate the blob $B(t_2)$. Upon joining the experiment, Q measures G's macrostate at t_2.

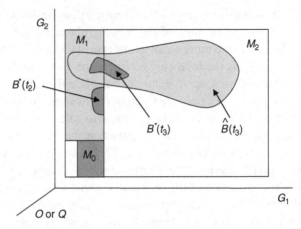

Figure 6.4 Dependence on observed history

Since the microstate of G is an objective feature of the world, O and Q agree on the macrostate of G at t_2: both say it is $[M_1]$. However, since for Q $[M_1]$ is the maximal information about G, Q will assume that the microstate of G can be anywhere in $[M_1]$ (and not only within the overlap with $B(t_2)$, of which Q is unaware). Taking $[M_1]$ as the initial macrostate at t_2, the dynamical blob that Q will calculate for the time t_3 is illustrated in Figure 6.4; we denote this blob by $\hat{B}(t_3)$. (In Figure 6.4, the evolution as seen by O is in a darker shade and the evolution as seen by Q is lighter in shade.) Therefore, when carrying out calculations regarding the probabilities of the macrostates that G can have at time t_3, Q will use the dynamical blob $\hat{B}(t_3)$ that evolves from the *entire* macrostate $[M_1]$ at t_2, and will apply the Probability Rule as follows:

$$P([M_i], t_3 | [M_1], t_2) = \mu(\hat{B}(t_3) \cap M_i)$$

By way of summing up this point, let us repeat and emphasize that although O and Q calculate different probabilistic predictions about G's macrostates, their predictions as well as the differences between their predictions are objective and physical. Observers O and Q have the same sort of physical interaction with G and are subject to similar constraints, and therefore they agree on the partitioning of G's state space into macrostates. And since G's microstate is an objective feature of the world, O and Q must always agree on G's macrostate. The only point that distinguishes O from Q is the *history* they observed: O knows more about the history of G than does Q, and therefore calculates a more precise description of the region in the state space of G that contains the actual microstate of G. Consequently, O calculates better predictions concerning G's evolution. This result features in explaining the spin echo experiments, to which we now turn.

6.7 The spin echo experiments

The spin echo experiments are an empirical example of the idea that observed history sometimes matters for probabilistic predictions of the evolution of physical systems. In these experiments, an observer such as O who knows the macroscopic history of the system more accurately than another observer Q will give more accurate predictions concerning the future evolution of the system. This is the way in which these experiments ought to be explained in statistical mechanics.

We begin with a brief description of the spin echo experiments. A sample of glycerin is placed for a while in a strong magnetic field in the z direction. This field affects the spins of the protons in the hydrogen atoms in the glycerin, and causes them to align in the z direction. At this stage the spin system emits a strong signal. Then, at time $t = 0$, an intense electromagnetic pulse with a suitable radio frequency is switched on for a short time interval, making the spins behave in a way that resembles spinning in the xy plane.[16] Following the application of the radio frequency pulse, the signal emitted by the spin system starts to decay until it practically vanishes at time $t = 1$ (where 1 stands for an appropriate time interval); see Figure 6.5. At $t = 1$ a second pulse is switched on for a duration twice as long as the first one. Consequently (as we will explain shortly), the intensity of the signal starts to increase, until at $t = 2$ an echo is obtained: the signal is only slightly weaker than the original signal; see Figure 6.5. At this stage the signal decays again, until at $t = 3$ another radio frequency pulse is applied, which results in an echo at $t = 4$, and so on. On repetitions of this process, the echo gradually decays until it completely fades away, and further applications of the radio pulse do not reproduce it.

Three features of the experiment call for explanation: (i) the disappearance of the signal in the short term ($t = 1, 3, 5$, etc. in Figure 6.5); (ii) the

---

[16] We use the term "resemble" to indicate that any talk about spinning here is a quasi-classical analog of an essentially quantum mechanical phenomenon. This analogy is intended to help in forming a figurative image of the experiment in the spirit of the papers of Hahn (1950, 1953).

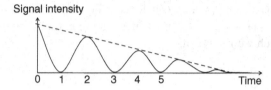

Figure 6.5 Schematic description of the intensity of the signal emitted during the spin echo experiment

return of the echo signal ($t = 2$, 4 etc. in Figure 6.5); and (iii) the long-term gradual decay of the signal's intensity from $t = 0$ to $t = 2$, from $t = 2$ to $t = 4$, etc. We begin with features (i) and (ii), and return to (iii) later.

Erwin L. Hahn, who conducted the first spin echo experiments in 1950, proposed the following analogy to explain the experiments:[17] the proton spins rotating in the xy plane are compared to a number of runners in a stadium in which they are initially all aligned. At $t = 0$ they start running, and because of the differences in their speeds, at time $t = 1$ (with some appropriate time unit) their positions along the track are different. At time $t = 1$ the runners turn around simultaneously, so that the last runner becomes first. If they run back with exactly the same speed that they had in the forward direction, at $t = 2$ the runners will be aligned along the original line.

Using the above analogy, Hahn offers the following explanation of the experiment, which is illustrated in Figures 6.6 and 6.7. At $t = 0$ the spins are all aligned in the same direction in the xy plane. This state is the initial macrostate of the system, denoted by $[M_0]$ in Figure 6.7. As usual, the initial macrostate is the basis for calculation of the dynamical blob, and is taken to be the blob at $t = 0$. Then, immediately after the first radio pulse is applied, the spins begin to rotate around the z axis. Some rotate faster, some slower; see dashed lines in Figure 6.6. As the spins become gradually unaligned the signal's intensity decreases. It would decrease by the same amount even if the details of the distribution of the velocities (and hence angles or positions) of different spins were different; in other words, the signal would not have changed if the *microstate* of the spins had been slightly different. In this sense, the intensity of the signal is an indication of the *macrostate* of the spins. As the dynamical blob evolves, it overlaps with different macrostates, corresponding to decreasing signal intensity, until at $t = 1$ the blob $B(t = 1)$ is within the macrostate $[M_1]$ in which the

[17] See Hahn (1950, 1953).

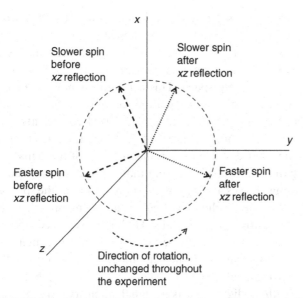

Figure 6.6 Schematic description of the evolution of the spins during the spin echo experiment

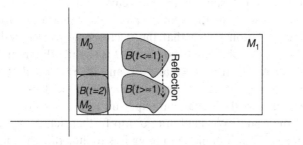

Figure 6.7 State space representation of the spin echo experiment

intensity of the signal is zero; see Figure 6.7. (For the sake of simplicity, the intermediate macrostates, corresponding to intermediate signal intensities, are not depicted in the figure. It might happen, in general, that the blob would partly overlap with other macrostates as well, with the appropriate probability distribution; we do not indicate this possibility here, since this is not the main point we wish to make.)

Then, at $t = 1$, a radio pulse is applied, and its effect on the spins is to *reflect* them relative to the xz plane. In Figure 6.6 this xz reflection is illustrated by the dotted lines which are the reflection of the dashed lines, relative to the xz plane. This reflection is analogous to the simultaneous reversal of velocity of the runners in Hahn's story; in other words, it is analogous to a velocity reversal, as explained in Chapter 4. In terms of state space, as illustrated in Figure 6.7, this reflection means that the dynamical blob is displaced in an *instantaneous* and *discontinuous* way, to a new region in the state space. This evolution is not dynamical in the sense that it does not follow the equations of motion of the spins, and it is brought about by an abrupt change of external constraints.[18] The displaced blob is still within macrostate $[M_1]$, since the reflection does not bring about an immediate change in the intensity of the signal.

Following the xz reflection, the spin that evolved fastest is now behind the others (using Hahn's analogy of the runners) and as the spins continue to rotate in the same individual frequencies as before, they gradually approach their initial aligned state. Had they returned to that state exactly, their macrostate would be $[M_0]$ again, and the intensity of the signal would be as strong as in the beginning. But experience tells us that the final macrostate of the spins at $t = 2$ is not $[M_0]$ but a different macrostate, with a slightly weaker signal intensity; this is macrostate $[M_2]$ in Figure 6.7. The gradual decrease of the intensity of the signal from the peak at $t = 0$ to the peaks at $t = 2, 4$ etc. is normally explained by the interaction of the spins with their environment: the environment induces perturbations on the spins, which take them away from their deterministic evolutions, such that the microscopic trajectory from the microstate at $t = 1$ to the microstate at $t = 2$ is slightly different from the microscopic trajectory from $t = 0$ to $t = 1$, and the result is that the spins do not return to their initial states simultaneously. The lack of simultaneity in the return to the initial state appears to us as a decrease in the intensity of the signal. We shall not go into further details of explaining this experiment;[19] our aim in mentioning it is to illustrate the importance of the observed history in calculating predictions in statistical mechanics, and it is to this that we now turn.

[18] Of course, such a change causes a change in the accessible region, and within this new accessible region the evolution is perfectly dynamical; and the removal of the pulse results in another change of the accessible region, namely a return to the previous one. For simplicity of exposition we do not depict this double change of accessible region, and express this event by the apparently discontinuous displacement of the dynamical blob near the time $t = 1$.

[19] For a quantum mechanical account of this effect in the spin echo experiment, see Hemmo and Shenker (2005). However, in order to explain the decay of the signal in both classical and quantum contexts, interventionist approaches require a choice of a probability measure over the environment degrees of freedom relative to which the perurbations are random. This choice needs to be justified. See more on this issue in Chapter 8.

Let us now compare the predictions of two observers O and Q, who predict the behavior of the spins at $t = 2$ on the basis of what they know about the spins at time $t = 1$. Observer O prepared the spins in their initial aligned state $[M_0]$ at $t = 0$. On the basis of this information, O can follow the evolution of the blob from $B(t = 0)$ to $B(t < \approx 1)$ and then apply the appropriate second radio pulse, predicting that it will bring about the externally induced displacement of the blob to $B(t > \approx 1)$ and from there predict the dynamical evolution to $B(t = 2)$.

Consider now the second observer Q who joins the experiment immediately after time $t = 1$, that is, right after the xz reflection. Since O and Q have the same kind of correlations with the spins, they agree on the partitioning of the state space of the spins to macrostates; and since the microstate of the spins is an objective feature of the world, Q agrees with O about the macrostate of the protons at this time, namely, that it is $[M_1]$. However, unlike O, Q does not know the history of the system at times earlier than $t = 1$, and therefore takes $[M_1]$ to be the initial macrostate, on which all predictions of transition probabilities are based, in accordance with the Probability Rule. In terms of Figure 6.4, Q will calculate the evolution of the entire region of $[M_1]$; $B(t_2)$ will, for Q, have the same Lebesgue measure as the measure of $[M_1]$, while for O the blob $B(t_2)$ had the same Lebesgue measure as $[M_0]$. Some of the trajectory segments that start out in $[M_1]$ at $t > \approx 1$ must end up in $[M_2]$; these are the trajectory segments that start out within the region B ($t > \approx 1$). But if macrostate $[M_1]$ happens to be larger (by the standard Lebesgue measure or another appropriate measure) than $[M_2]$, then (according to Liouville's theorem) there must be trajectory segments that start out in $[M_1]$ at $t > \approx 1$, and that reach other macrostates at $t = 2$. The observer Q will calculate the probabilities of those other macrostates, in accordance with the Probability Rule. Thus, the probability that the spins will end up at $t = 2$ in $[M_2]$ assigned by Q will in general not be 1. This means that the probability of $[M_2]$ at $t = 2$ as calculated by Q will differ from the probability calculated by O for the same event.

It is often assumed that, taking $[M_1]$ as the initial macrostate, Q would assign a very low probability for $[M_2]$ at $t = 2$, and a high probability for $[M_1]$ at $t = 2$, and so conclude that the spins are very likely to remain in $[M_1]$ after $t = 1$. Such an assumption is based on an argument analogous to Boltzmann's combinatorial construction of macrostates, discussed in Chapter 5. If indeed the probability that Q assigns for the macrostate $[M_2]$ at $t = 2$ is small, we may reasonably guess that Q will be quite surprised to see that O can bring the spins to that macrostate time and again, at will.[20] But this will be surprising only for Q who is not aware of O's information. Since we know the observed history of O, we do not take this fact to be at all surprising.

However, the spin echo experiments are still unique, and it is import- ant to see why and what exactly is unique about them; for one needs to explain not only what happens in this special case, but also what happens in the usual cases (which we describe in the next section). And so, in the context of the spin echo experiments, it is important to distinguish between two notions that are often closely linked together: *information* and *control*. The uniqueness of the spin echo experiments lies in the fact that in these experiments the link between information and control is broken, in the following sense. By comparison, think of a precise and instantaneous reversal of the velocities of all the particles of a gas (called Loschmidt reversal; see Chapter 7). To carry it out we would need precise information about the velocities of all the gas particles in order to place tiny walls at the right positions and angles so that the velocities would be exactly reversed. If our information is not precise, the reversal will not be precise, and then, because of the (supposedly) complex nature of the dynamics of the particles, the gas will not trace back its evolution. Hahn's runners perform such a reversal: they all hear the signal and turn around simultaneously, and follow their exact original trajectories from the end to the start. Here, the signal is a sort of macroscopic manipulation that works with no need to know the different velocities of the individual runners. In other words, the control over the runners' behavior exceeds the information about them. Can we carry out such a macroscopic manipulation in the gas, one that will bring about a Loschmidt reversal in all its particles? As things stand now, the only case in which we know how to do this is the spin echo experiment: only here can a macroscopic signal, which is in this case the right sort of radio pulse, bring about a microscopic manipulation analogous to a precise Loschmidt reversal, with no need to know the microscopic state of the system. In other words, in the spin echo case our control exceeds the amount of information given by the macrostate of the spins.[21]

---

[20] Ridderbos and Redhead (1998) take this surprise as one argument for rejecting the idea of coarse-graining in Gibbsian statistical mechanics; we discuss Gibbs's ideas in Chapter 11.

[21] Note that this explanation makes no reference to notions such as *quasi*-equilibrium; see Blatt (1959).

6.8 Robustness of transition probabilities

It is a fact that the thermodynamic regularities are *robust* in the sense that they are independent of observed history. Indeed, this is what makes the spin echo experiments so unique. On the other hand, according to the Probability Rule the general case is that probabilities of macrostates do depend on observed history. For this reason the thermodynamic regularities call for explanation. In this section we show how to express this robustness of the transition probabilities or independence of history in terms of the Probability Rule. The question of whether mechanics entails that the thermodynamic regularities are indeed the case in our world is the topic of Chapter 7.

Consider again the two observers O and Q (from Section 6.6) who differ only in their observed history (see Figure 6.4). Observer O has prepared the system in the macrostate $[M_0]$ at t_1. Observer Q has no knowledge about this preparation. At t_2 both O and Q find the system in the macrostate $[M_1]$. As we saw, at t_2 O calculates the probabilities for the macrostates at t_3 on the basis of the blob $B^*(t_3)$, while Q calculates the probabilities for the macrostates at t_3 on the basis of the blob $\hat{B}(t_3)$. Observer O assigns probabilities to any macrostate $[M_i]$ at t_3 given $[M_0]$ at t_1 by $P(M_i, t_3 | M_0, t_1) \approx a_i$, while Q assigns the probabilities to any macrostate $[M_i]$ at t_3 given $[M_1]$ at t_2 by $P(M_i, t_3 | M_1, t_2) \approx b_i$. We can express the fact that the thermodynamic regularities are independent of observed history as the idea that $a_i = b_i$. This immediately implies that $\mu(\alpha) : \mu(\beta) = \mu(\gamma) : \mu(\delta)$ (see Figure 6.8). A very special case, which may or may not hold in thermodynamic evolutions, is that in which the probability of a macrostate is equal to its measure: that is, $\mu(\alpha) : \mu(\beta) = \mu(\gamma) : \mu(\delta) = \mu(M_1) : \mu(M_2)$.[22]

And so given that, as a matter of fact, the thermodynamic regularities are independent of observed history, it turns out that the underlying dynamics has this feature. And here we see how the interplay between blobs and macrostates allows us to infer from the thermodynamic experience some specific conclusions about the structure of trajectories of thermodynamic systems.

6.9 No probability over initial conditions

Our treatment of probability in statistical mechanics is different from those that one usually finds in the philosophical literature on the subject.

---

[22] The μ-measure of a macrostate may be different from entropy; see Chapter 7.

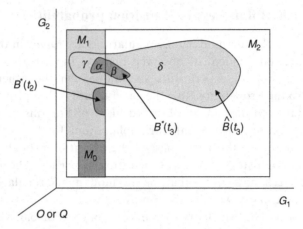

Figure 6.8 Dependence on observed history

It is instructive to discuss this difference. For example, in his book *Time and Chance* David Albert describes what he calls the *Statistical Postulate* of statistical mechanics as follows:

> The right probability distribution to use for making inferences about the past and the future is the one that's uniform, on the standard measure, over those regions of state space which are compatible with whatever other information – either in the forms of laws or in the form of contingent empirical facts – we happen to have.[23]

Albert's picture of statistical mechanics is essentially this. We start by assigning a macrostate $[M_0]$ to the system in question at at a time t_0. This macrostate expresses everything we know about the system's state at that time. Albert's statistical postulate *adds* that the actual microstate of the system is equally likely to be anywhere within that macrostate; more precisely, it is just as likely to be in subregions of this macrostates which have the same Lebesgue measure (which is the standard measure mentioned by Albert). According to Albert, this statistical postulate together with the Newtonian laws of motion yields the empirical regularities that we observe in thermodynamics.

However, our Probability Rule does not refer to nor does it entail anything about the probability distribution over the microstates in $[M_0]$.

---

[23] Albert (2000, Ch. 4, p. 96). Albert attempts to justify his statistical postulate by appealing to Loewer's (2001) approach according to which the statistical postulate receives the status of a law of nature. This of course presupposes the general validity of the laws of thermodynamics which we question in Chapters 7, 8, and 13.

And so how does Albert's idea connect with our above proposal for calculating probabilities on the basis of overlaps between dynamical blobs and macrostates? We will now show that in many interesting cases Albert's proposal and ours yield similar predictions. However, the explanation and justification for these predictions are very different in the two approaches.

We begin by illustrating the differences between the approaches, using again the example of a gas expanding in a container. We prepare a large sample of similar systems, in all of which at time t_0 the gas is confined to a part of the container by a partition, and in all of which the macrostate of the gas at this time is $[M_0]$. At t_1 the partition is removed and the gas is allowed to expand in the container. Figure 6.2 is the state space description of this experiment. Suppose that we find that at time t_2 1/3 of the gases in our sample arrive at the macrostate $[M_1]$ and 2/3 of them arrive at $[M_2]$. According to our proposal, as described in the previous sections of this chapter, we proceed to apply the Probability Rule in order to calculate the probabilities that the next gas, prepared in $[M_0]$, will arrive at $[M_1]$ or $[M_2]$. We choose a measure μ such that the overlaps of $B(t_2)$ with $[M_1]$ and $[M_2]$ have the measures 1/3 and 2/3, respectively. There is no other way to justify the use of probability measure in statistical mechanics (see more on this issue in Chapter 8).

But now consider how, and to what extent, our way of assigning probability can fit Albert's statistical postulate, in some interesting cases. Consider again the case illustrated in Figure 6.2. Suppose that, as a matter of contingent fact, it turns out that the *Lebesgue* measures of the overlaps between the blob $B(t_2)$ and the macrostates $[M_1]$ and $[M_2]$ match our observations; and since the measure is doubtless very convenient, we take it to be the appropriate measure for calculating probabilities for this system. Now, one of the properties of the Lebesgue measure that makes it so convenient is that it is conserved under the dynamics of classical mechanics. For example, in this case, if we take the set of all the microstate points within $[M_0]$ which lead to $[M_1]$ at t_2, we will find that its Lebesgue measure is exactly equal to the measure of the overlap between $B(t_2)$ and $[M_1]$. A similar outcome will be obtained for all dynamical blobs that evolve from $[M_0]$, at all times. And so in this case, if one starts with Albert's statistical postulate, and postulates a uniform probability distribution relative to the Lebesgue measure over $[M_0]$ at t_0 together with the equations of motion, then one predicts that, for example, the probability that the gas will arrive at $[M_1]$ at t_2 is 1/3. In this case Albert's predictions would coincide with ours.

What, however, explains the success of Albert's predictions? What seems to be a prediction based on probability distribution over initial conditions is fundamentally based on *retrodiction*. Furthermore, as we have just explained, the very choice of the Lebesgue measure, as the measure relative to which probability is distributed over $[M_0]$ at t_0, is based on the fact that the Lebesgue measures happened to yield $1/3$ as the measure of the overlap of $B(t_2)$ and $[M_1]$, and this number happens to match our experience of relative frequencies. This means that the success of Albert's predictions can be explained only on the basis of the Probability Rule and only if the measure μ is the Lebesgue measure.

Moreover, whether or not the μ-measure turns out to be the Lebesgue measure is a contingent matter of fact: for all we know, it may happen that the Lebesgue measure of the overlap of $B(t_2)$ and $[M_1]$ would yield a number that does *not* match the observed relative frequencies. In that case we would have to take a different measure as μ in the Probability Rule. And there is no guarantee that the chosen measure would be conserved under the dynamics of classical mechanics; so there is no guarantee that the measure of the overlap of $B(t_2)$ and $[M_1]$ would be the same as the measure of the points within $[M_0]$ that evolve to this overlap. Thus our probabilistic predictions would be different from those of Albert.

7

Entropy

7.1 Introduction

In this chapter we begin to tackle head-on the problem of the direction of time. As we saw in Chapter 2, the thermodynamic regularities seem to provide an arrow of time since they are *time-asymmetric*: the Law of Approach to Equilibrium describes a universal approach to equilibrium as we move from past to future, and the Second Law describes an increase of entropy towards the future, as constraints are removed. By contrast, as we saw in Chapters 3 and 4, classical mechanics is *time-symmetric*. And so, if one attempts to underwrite the thermodynamic regularities by classical mechanics, one has to explain how time asymmetry arises in a mechanical time-symmetric world.

What would it take to underwrite the time-asymmetric thermodynamic regularities on the basis of the time-symmetric classical mechanics? In this chapter we apply the ideas presented in the previous chapters in order to answer this question. It is crucial to emphasize from the outset that we are not going to provide mechanical *proofs* of the thermodynamic regularities. Indeed, in our view it is no wonder that such proofs have not been given by anyone so far, despite great efforts that began with Boltzmann; and it is one of the main objectives of this book to explain why it is that such proofs are not going to be given, indeed why such proofs cannot be given, except perhaps for very special and limited cases. To say that such proofs are impossible is to say that the thermodynamic regularities are not universal, not even in their probabilistic versions; and indeed this is what we are going to show (in Chapter 13) by the counter-example known as Maxwell's Demon, to which our journey in this book leads. In view of all this, the objective of the present chapter is to express, in terms of blobs and macrostates, the thermodynamic regularities of the Second Law and the Law of Approach to Equilibrium, and, once these laws are couched in mechanical terms, to see what it would take to prove them in statistical mechanics.

The central thermodynamic notions involved in the time asymmetry that characterizes this theory are equilibrium and entropy. We have already said that an important part of underwriting thermodynamics by mechanics is finding mechanical counterparts of the thermodynamic magnitudes, and that these counterparts are properties of macrostates. Thermodynamic states correspond to macrostates, and thermodynamic magnitudes correspond to properties or shared aspects of the microstates belonging to macrostates. We already saw, in a general outline, how the thermodynamic magnitudes such as pressure, volume and temperature are translated into mechanical magnitudes: roughly, volume corresponds to the distribution of the positions of the particles; temperature is a function of the velocities of the particles; and pressure depends on both. When the mechanical degrees of freedom, of positions and velocities, are partitioned into sets, we get mechanical macrostates which correspond to different values of volume, temperature, and pressure. To describe the thermodynamic regularities in terms of statistical mechanics, we now need to consider the way in which the notion of thermodynamic entropy is to be understood as a property of macrostates. We begin with the notion of entropy.

7.2 Entropy

In thermodynamics, entropy designates the degree to which the energy of a system is exploitable in order to produce work. In the passage from thermodynamics to mechanics, the notion of exploitability of energy translates into the notion of the degree of control that one has over the energy in a given system. If one can control the way in which energy flows in a system and changes its form, one can, naturally, exploit this energy more easily. In thermodynamics, the convention is to associate *low* entropy with *high* exploitability of energy, and *high* entropy with *low* exploitability of energy. In statistical mechanics this convention is preserved: low entropy is associated with high control over the system's energy, and high entropy denotes low control over energy. How can the notion of control be expressed in terms of statistical mechanics?

In mechanics, the real physical state of a system is its microstate, which is represented in the state space by a mathematical point. Since the positions and velocities of particles are real numbers that can have any value in a given interval of the continuum, describing the exact microstate requires an infinite amount of information about the system's microstate, and this is practically impossible. Because the degree of precision with which we can know the state of the system is finite, this means that we cannot determine the system's exact microstate but can – at best – determine a set of microstates to which the system's microstate belongs; in other words, we can only determine its macrostate. And if our knowledge of the system's state is limited in this way, then *ipso facto* so is our ability to control this state (with the exception of the spin echo experiments). The result is that we cannot prepare a system in a predetermined microstate but – at best – prepare it in a predetermined macrostate. Macrostates express, as a rule, not only our limited *information* about a system's actual microstates, but also our limited *control* over its microstate. Figuratively speaking, the borderlines of the macrostates are the closest we can get to the actual but unknown microstate of the system; they are the extent to which we can approach the system's microstate and control it. And so, the smaller the macrostate is, the closer we can get to the actual microstate of the system, and the better is our control of its actual microstate.

Now, the energy of a system is a function of the system's microstate; and therefore in order to control the energy of a system and exploit it in a useful way, one needs to determine and control the system's microstate. And since the degree of control over the microstate is associated with the size of that macrostate (as we have explained), it turns out that the larger the system's macrostate, the lower is the degree to which we can exploit its energy in a useful way. Going back to the idea of entropy in thermodynamics, and recalling that high entropy is associated with low degree of exploitability of its energy, it turns out to be reasonable to associate the size of a macrostate with the entropy of a system, which is in this macrostate. The notion of entropy of a macrostate in mechanics is naturally given in terms of the *size* of the macrostate in the state space.[1] This is,

---

[1] Reichenbach (1956, pp. 61–64) agrees that taking W in Boltzmann's expression to be the measure of the macrostate best reflects the meaning of entropy as a state function in thermodynamics. He points out, however, that in the framework of Boltzmann's approach, in which the probability of a macrostate is relative to the normalized entropy of that macrostate, this leads to an analog of Gibbs's paradox. Reichenbach's solution is "repeated regauging": "If a thermodynamical system goes from the entropy S_1 to the entropy S_2, and then from S_2 to S_3, the physicist coordinates a probability increase to the transition from S_1 to S_2 and another probability increase to the transition from S_2 to S_3; but these probability increases are not conjoined into one increase." The problem that

indeed, the standard notion of entropy in statistical mechanics following Boltzmann's definition of the entropy of a macrostate M:

$$S_M = k \log W,$$

where k is the so-called Boltzmann's constant (which translates units of thermodynamic temperature into the mechanical units of energy) and W is the size of that macrostate.[2] This expression is one of Boltzmann's greatest achievements.[3]

We saw, however, in our earlier discussions about probability, that there is no unique measure that can be used to determine the size of a region in the state space. It is customary to use the Lebesgue measure here, and the question now is this: what justifies the choice of this particular measure in determining the entropy of a system in a given macrostate? It is to this question that we now turn.

When a measure is chosen in order to calculate the transition probability from an initial macrostate $[M_0]$ to a given macrostate $[M]$ after a given time interval t, according to the Probability Rule (see Chapter 6), the measure μ is chosen on the *empirical* basis of observed transition probabilities. We would like the choice of the measure of entropy to be based on empirical considerations as well. Moreover, and perhaps more important, we would like the choice of the entropy measure to be such that a natural link will be formed between the statistical mechanical notion of entropy and the notion of entropy in thermodynamics. We will now argue that if this choice is indeed based on empirical considerations which are relevant to thermodynamics, then the measure μ in the Probability Rule

---

bothers Reichenbach does not arise in the framework put forward in this book, where the probability of a macrostate is independent of the entropy of that macrostate.

[2] One reason for taking the logarithm here, rather than the simpler expression kW, is that (in general outline) in thermodynamics, entropy is an additive quantity. That is, if one brings together two gases, one with entropy x and the other with entropy y, the entropy of the two-gas system will be $x + y$. Now in statistical mechanics, if one combines two gases, one creates a new state space consisting of the degrees of freedom of both of them; and the volume is the multiplication of the original volumes: xy. Thus additivity is lost. But the logarithm of the volumes is still additive: $\log(xy) = \log(x) + \log(y)$, and therefore we use the logarithm in the expression for entropy, in order to retain the additivity which is characteristic of thermodynamic entropy.

[3] The magnitude W in Boltzmann's expression for entropy is often read as the probability of the macrostate. This would be the case if the size of a macrostate and its probability were the same, or at least a function of each other. But we will see that this is not the case. For a discussion of the various interpretations of W, see, for example, Swendsen (2011). On Boltzmann's ideas concerning W, see, for example, Uffink (2008). For a comparison of Botzmann's notion of entropy with the Gibbsian one, see Jaynes (1965).

and the measure of entropy we are looking for, which we will denote by v, need not in general be the same.

Take, for example, the thermodynamic entropy difference between two equilibrium states A and B of an ideal gas, which is given by the following expression:

$$\Delta S = \int \frac{\delta Q}{T} = c_V \ln\frac{T_B}{T_A} + R \ln\frac{V_B}{V_A},$$

where T and V are the temperature and the volume of the gas, and c_V and R are constants.[4] The quantity ΔS is a way to compare two equilibrium states (corresponding to two sets of external constraints), and this comparison is one of the main achievements of the formulation of the Second Law of Thermodynamics in terms of entropy. The task now is to find an expression in statistical mechanics that would make a comparison between equilibrium states in a way that parallels the thermodynamic comparison between equilibrium states; and that would, therefore, be the basis for a recovery of the Second Law of Thermodynamics within statistical mechanics.

The temperatures T_A and T_B of the gas are associated with the macrostates $[T_A]$ and $[T_B]$, which differ in the velocity degrees of freedom of the gas particles; and the volumes V_A and V_B are associated with the macrostates $[V_A]$ and $[V_B]$ which differ along the position degree of freedom of the particles. Combining all the degrees of freedom together, when the gas is in the thermodynamic states A and B, its microstates are within the macrostates $[M_A]$ and $[M_B]$, respectively. We know from the above considerations that the size of the regions $[M_A]$ and $[M_B]$ (according to some measure) corresponds to the thermodynamic entropy of the gas when it is in these macrostates. The question is now this: which measure is most suitable for this purpose? In order for our measure to be such that thermodynamics will be recovered, in the special case of an ideal gas, we now look for some measure v over the state space, so that the following equation will be satisfied:

$$\Delta S = k \ln\frac{v([M_A])}{v([M_B])} = c_V \ln\frac{T_B}{T_A} + R \ln\frac{V_B}{V_A}.$$

Here, we are looking for a quantitative equality between expressions, one of which belongs to thermodynamics and the other of which belongs to

---

[4] For a derivation of this expression, see any introduction to thermodynamics, for example Fermi (1936).

statistical mechanics. Any measure v that satisfies this condition is acceptable. Assuming that more than one measure can satisfy this requirement, we will choose the one that is the most convenient, just as we did in the case of the probability measure μ in the Transition Probability Rule. We will then take the logarithm of the v-measure of a macrostate in the state space as corresponding to thermodynamic entropy.

At this point it is important to see a significant difference between the thermodynamic entropy and the statistical mechanical entropy. Standard thermodynamics ascribes magnitudes only to equilibrium states, and accordingly, entropy difference is defined only between two equilibrium states. However, once we define the statistical mechanical magnitude corresponding to entropy as the v-measure of an equilibrium macrostate, then there is, *prima facie*, no reason to restrict the definition to equilibrium macrostates, and we can therefore generalize our result and take the measure v of *any* macrostate as designating its entropy. (See Chapter 11 for some further ways of justifying this move.)

7.3 The distinction between entropy and probability

The transition probability from an initial macrostate $[M_0]$ at time t_0 to some macrostate $[M]$ at some later time t, is the measure μ of the overlap between $[M]$ and the dynamical blob $B(t)$ that started out in $[M_0]$ at time t_1; whereas the entropy of $[M]$ is the v-measure of $[M]$ itself, and has nothing to do with either $[M_0]$ or $B(t)$. And so there is no reason to expect that the *probability* of $[M]$ will have anything to do with the entropy of $[M]$.

This realization deviates significantly from a mainstream view on the subject, and hence ought to be emphasized. It is a consequence of our introduction of the notion of a dynamical blob and of the idea that the overlaps between dynamical blobs and macrostates are the dynamical origin of probability in statistical mechanics.

It may happen that once the measure v is chosen as explained above, it will turn out to be the same as the measure μ in the Probability Rule. But while there is no reason to rule out such a possibility, there is no reason to think it is general, and even less reason to require it in selecting the two measures. Our reasons for choosing v are generally quite different from our reasons for choosing the measure μ in the formulation of the Probability Rule.

At the same time, there is an important linkage between the probabilities as determined by the measure μ of the overlap between blobs and macrostates, and the entropy as determined by the measure v of macro-states: the linkage is via the probabilistic counterparts of the Law of Approach to Equilibrium and the Second Law of Thermodynamics. The idea is that the evolution of thermodynamic systems happens to be – as a matter of fact (expressed by the *empirical generalizations* of thermodynamics) – such that it is highly probable that the entropy of macrostates will satisfy these laws. We now turn to see this linkage in more detail.

7.4 Equilibrium in statistical mechanics

The first stage in formulating the statistical mechanical counterparts of the laws of thermodynamics is to formulate a statistical mechanical notion of equilibrium. In thermodynamics, recall that equilibrium is a state in which the thermodynamic magnitudes are constant over time, and the laws and magnitudes of standard classical thermodynamics are defined for such states only. The reason behind this is that in order for a measuring device to record the state of the measured system, it has to arrive at some sort of equilibrium with that system, so that the value recorded stably reflects a property of the measured system. The idea that equilibrium is a stable state should be expressed in mechanical terms as a property of the equilibrium macrostate.[5]

In statistical mechanics *equilibrium is a macrostate*: when a system's microstate is in this macrostate we shall say that it is in equilibrium. The equilibrium macrostate $[M_{eq}]$ is best characterized in one of the following two ways.

(I) If a system is already in $[M_{eq}]$ then the probability of staying in it during a time interval from t to $t + \Delta t$ is the highest, relative to the probability of staying in any other macrostate during that time interval from t to $t + \Delta t$. The probability for staying in a macrostate $[M_n]$ during the time interval from t to $t + \Delta t$ is given by the overlap of the dynamical blob, that starts out in $[M_n]$ at t, with $[M_n]$ at $t + \Delta t$. Formally:

$$P(M_n(t + \Delta t)|M_n(t)) = B(t + \Delta t) \cap M_n,$$

---

[5] A different notion of equilibrium is given in the Gibbsian approach. We discuss this approach in Chapter 11.

where $B(t + \Delta t)$ is the blob that starts in $[M_n]$ at t and evolves to $t + \Delta t$ according to the equations of motion of the system.

(II) Given some initial macrostate $[M_0]$, the transition probability to the macrostate $[M_{eq}]$ generally increases with time, until it reaches a stable maximum.

Note that the equilibrium macrostate is not defined as a macrostate of high *entropy* (see Chapter 2). This is a special case of our distinction between the probability of a macrostate and the entropy associated with that macrostate.

7.5 Law of Approach to Equilibrium

We start by analyzing the way the Law of Approach to Equilibrium would look in statistical mechanics, that is, in terms of macrostates, dynamical blobs and transition probabilities, together with the measure μ of probability and the measure ν of entropy; and where the notion of equilibrium is as described above. (Recall that we will *not* prove this law, because it is generally false, as we show later in this book.)

Consider Figure 7.1, which gives a simplified state-space description of an approach to equilibrium. As an illustration, think of a gas G expanding in a container, as in Figure 1.1, where G_1 and G_2 in the figure are sets of degrees of freedom of the gas, which qualitatively represent the general behavior of the gas in the state space. The gas is prepared in macrostate

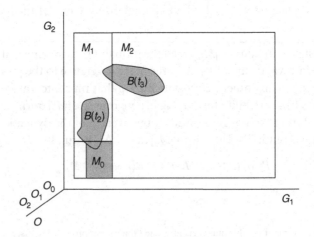

Figure 7.1 The Law of Approach to Equilibrium

$[M_0]$, at which it starts out at time t_1. The dynamical evolution of the gas is such that at time t_2 the dynamical blob is spread over macrostates $[M_0]$ and $[M_1]$, and the transition probabilities from $[M_0]$ to either $[M_0]$ or $[M_1]$ is given, according to the Probability Rule, by the measure μ of the overlaps between the blob $B(t_2)$ and these macrostates, in accordance with the observed relative frequencies of these macrostates in similar experiments. Notice that we assume, in this example, that at t_2 the overlap of the blob $B(t_2)$ with the macrostate $[M_2]$ is empty, i.e. the transition probability from $[M_0]$ at t_1 to $[M_2]$ at t_2 is zero. This zero probability has here the meaning of strict impossibility: no trajectory that starts out in $[M_0]$ at t_1 reaches $[M_2]$ at t_2. (This is not a general claim, but we focus on it in order to emphasize that in general not all the macrostates can be accessed within a short enough time interval after the initial time.) After t_2 the dynamical blob continues to evolve such that at time t_3 the blob $B(t_3)$ is spread over $[M_1]$ and $[M_2]$ (but not over $[M_0]$, in our example), and the transition probabilities from $[M_0]$ at t_1 to $[M_1]$ and $[M_2]$ at t_3 are again given by the Probability Rule.

Notice that when we described both the blob $B(t_2)$ at t_2 and the blob $B(t_3)$ at t_3, we thought of each of them as evolving from the initial macrostate $[M_0]$ at t_1. In particular, $B(t_3)$ was described as evolving from $[M_0]$ and not from $B(t_2)$. The reason is that the Probability Rule is about predicting future macrostates given the initial macrostate. We always describe evolutions in terms of *macrostate to macrostate*; the dynamical blob is merely a computational tool, and is not directly observable. This fact will be important in formulating the statistical mechanical counterpart of the Law of Approach to Equilibrium.

For a probabilistic counterpart of the Law of Approach to Equilibrium to be true in the above case, we would need the following conditions to hold: (I) it is highly probable that the gas G will evolve from $[M_0]$ through some intermediate macrostates such as $[M_1]$ to the macrostate $[M_2]$, and (II) once the system arrives at $[M_2]$, it is highly likely to remain there for all time. (III) To these two conditions we may add the requirement that $[M_2]$ has the highest entropy in terms of the entropy measure ν. Conditions (II) and (III) are conceptually distinct: entropy and probability are not the same, as we shall emphasize throughout this chapter. However, since the case in which equilibrium also has high entropy is an important one in thermodynamics, it is worth mentioning here. More formally, these three conditions are expressed as follows in terms of Figure 7.1:

Condition (I) $\mu(B(t_2) \cap [M_1]) >> \mu(B(t_2) \cap [M_0])$ and
$\mu(B(t_3) \cap [M_2]) >> \mu(B(t_3) \cap [M_1])$,

where $B(t_2)$ and $B(t_3)$ are the dynamical blobs that evolve from $[M_0]$ at t_1.

Condition (II) For all $t > t_3$ and for all $[M] \neq [M_2]$,
$\mu(B(t) \cap [M_2]) >> \mu(B(t) \cap [M])$.

Condition (III) $v([M_0]) \leq v([M_1]) \leq v([M_2])$.

These three conditions are formulated for the example above, but the generalization is straightforward: the Law of Approach to Equilibrium holds if the evolution is such that the trajectory of the system has high probability (as given by the Probability Rule with measure μ) of passing through a sequence of macrostates with *increasing entropies* (as given by measure v) until it reaches the macrostate with the highest entropy, called the equilibrium macrostate, in which the system is highly likely to remain at all times.[6]

In order to prove this law for a chosen v, one would need to show that the *dynamics* of the system in question, i.e. the evolution of its dynamical blob, satisfies the above conditions. On the one hand, there cannot be universal theorems proving this law, and the reason is that this law has a counter-example in the form known as Maxwell's Demon, which we discuss in Chapter 13. On the other hand, we see around us systems that do behave in accordance with the Law of Approach to Equilibrium, and therefore it is reasonable to expect that under suitable conditions, corresponding to the circumstances of the universe around us, a statistical mechanical version of this law can be proved. Indeed, there are some theorems to this effect concerning very special cases, such as Lanford's theorem, which we discuss below.

7.6 Second Law of Thermodynamics

In thermodynamics, the Second Law states that if the external constraints on a system in equilibrium are weakened, the system will approach a new equilibrium state that has the same or higher entropy. Consider, for example, the case illustrated in Figure 1.1: a gas confined to the left-hand side of a container by a partition. Suppose that at time t_0 the partition is removed, and the gas expands in accordance with the Law of Approach

---

[6] Compare with the Goldilocks mixing dynamics proposed by John Earman (2006, p. 406).

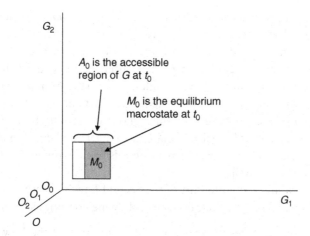

Figure 7.2 The Second Law of Thermodynamics: the earlier equilibrium state

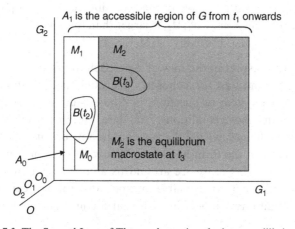

Figure 7.3 The Second Law of Thermodynamics: the later equilibrium state

to Equilibrium, and ends spread over the entire container. An outline of the state-space representation of this evolution is given in Figures 7.2 and 7.3 (compare Figures 6.1 and 6.2).

At time t_0, the system is constrained by the partition, and this external constraint determines the accessible region of the system to be A_0. Within the accessible region A_0 there are several macrostates, one of which is $[M_0]$. Assuming that at t_0 the system satisfies the Law of Approach to Equilibrium, we may assume that $[M_0]$ is the equilibrium state, relative to the accessible region A_0. At time t_1 the constraints on the system are relaxed (by removing the partition), and this results in an *instantaneous* increase in the volume of the accessible region, from A_0 to A_1. The new accessible region A_1 contains all the microstates that are consistent with the new, weaker constraints. Of course, at t_1, right after the removal of the partition, the gas is still in $[M_0]$; but while $[M_0]$ is still its macrostate, its accessible region is now A_1. The system that starts in any of the microstates within $[M_0]$ can now evolve throughout the region A_1. In general, once A_1 is the accessible region, $[M_0]$ is no longer the equilibrium state, and a new macrostate will take its place in this role. In Figure 7.3 the new equilibrium state is $[M_2]$.

For a probabilistic counterpart of the Second Law of Thermodynamics to hold in this case, we would need the following conditions to hold:

(I) The system satisfies the probabilistic counterpart of the Law of Approach to Equilibrium in the above sense, where the equilibrium macrostate relative to the accessible region A_0 is $[M_0]$ and the equilibrium macrostate relative to the accessible region A_1 is $[M_2]$.

(II) The entropy measure v of $[M_2]$ is larger than (or equal to) the entropy measure v of $[M_0]$. Formally: $v([M_0]) \leq v([M_2])$.

As we said with respect to the Law of Approach to Equilibrium, in order to prove the Second Law for a chosen v one would need to show that the dynamics of the system in question, i.e. the evolution of its dynamical blob, satisfies the above conditions. And here, too, there are no universal theorems proving this law, and there cannot be any, because this law has a counter-example in the form known as Maxwell's Demon. On the other hand, as we said above, we see around us ample cases that satisfy the Second Law, and so it is reasonable to expect that this law can be proven for special conditions that hold in the actual circumstances of the universe around us.

7.7 Boltzmann's *H*-theorem

The most famous attempt to prove that the dynamics of classical mechanics gives rise to the Law of Approach to Equilibrium is what came to be known as *Boltzmann's equation* and Boltzmann's *H-theorem*.[7] Boltzmann tried to show that this law is universally valid: that systems will invariably

---

[7] For more details, see Uffink (2004); Uffink (2007); Brown, Myrvold and Uffink (2009); Uffink and Valente (2010);

evolve to equilibrium, regardless of their initial microstate. Although, as is well known, Boltzmann's argument turned out to be incompatible with mechanics, it instructive to consider it, since we think that Boltzmann was fundamentally right in insisting that the matter should be addressed by dynamical considerations (as opposed to non-dynamical considerations of the kind that came later and are discussed below). We will now present Boltzmann's equation and *H*-theorem in terms of macrostates and blobs. This is in line with our characterization of the Maxwell–Boltzmann (MB) energy distribution as a macrostate in Chapter 5. Although this presentation is different from the standard formulations, we shall see that it has some advantages, and we shall explain how the two formulations are related.

Boltzmann's equation describes the evolution in time of a microstate of a gas, where this microstate is characterized by a certain distribution $[f(q,p)]$ of the positions q and momenta p of the gas particles. However, since any finite observer can distinguish only between intervals of continuous magnitudes such as position and momentum, and since permutations of particles of the same kind are presumably indistinguishable, in our terminology $[f(q,p)]$ is actually a macrostate (and therefore we put it in square brackets). (Indeed, $[f(q,p)]$ seems to be a macrostate of the kind that was later proposed by Boltzmann as part of his combinatorial argument; recall that we have cast this idea of Boltzmann in our terms in Section 5.6 and 5.7.)

In our terminology, what Boltzmann tried to do is to describe the evolution of the dynamical blob that starts out in a macrostates of the form $[f(q,p)]$; we denote the initial macrostate by $[f_0(q,p)]$. The equation describes the evolution of the dynamical blob that starts out in $[f_0(q,p)]$, and a *solution* of the equation would give the macrostate $[f_t(q,p)]$ which fully contains (or fully overlaps with) the blob for any time t after $t = 0$; the *full* containment is essential since Boltzmann attempted to give an absolute counterpart of the Law of Approach to Equilibrium, not a probabilistic counterpart. (We do not describe Boltzmann's equation in all its details.[8]) Boltzmann hoped to be able to show that after a time interval (unspecified but presumably finite and of the order of magnitude of the thermodynamic relaxation time) the blob enters the macrostate $[f^{MB}(q,p)]$, in which the positions of the gas molecules are distributed uniformly over the container and the momenta of the molecules are

---

[8] For details, see Uffink (2004, 2007).

distributed according to the Maxwell–Boltzmann distribution, which was already known to be characteristic of equilibrium; and that once the blob enters that macrostate it remains there indefinitely. Hence $[f^{MB}(q,p)]$ is to be identified as the equilibrium macrostate.

The challenge is, of course, to solve the equation and thus prove this result. We have already noticed, however, in Chapter 6, that calculating the evolution of dynamical blobs is extremely difficult, if not utterly unfeasible. Boltzmann was well aware of this difficulty and found a way around it, avoiding the necessity to solve his equation. His method around the need to solve the equation is known as the *H-theorem*.

Boltzmann defined the following quantity:

$$H([f(q,p)]) = \int [f(q,p)] \log [f(q,p)] \mathrm{d}q \mathrm{d}p,$$

where q and p stand for the distribution of the positions and momenta of the gas particles in the macrostate $[f(q,p)]$, and showed that the time evolution of H satisfies the inequality

$$\frac{\mathrm{d}H([f_t(q,p)])}{\mathrm{d}t} \leq 0,$$

where equality is obtained for the special macrostate $[f^{MB}(q,p)]$ which is characterized by uniform distribution in positions q and the Maxwell–Boltzmann distribution in momenta p. This is the *H*-theorem.

Boltzmann then associated the quantity $-H([f(q,p)])$ with thermodynamic entropy, and interpreted his *H*-theorem as underwriting the Law of Approach to Equilibrium. This law, in terms of Boltzmann's *H*-theorem, says that gases will universally end up in the Maxwell–Boltzmann distribution, which maximizes the entropy $-H([f(q,p)])$, and remain there indefinitely.

Boltzmann concludes his proof as follows. "It has thus been rigorously proved that whatever may have been the initial distribution of kinetic energy, in the course of time it must necessarily approach the form found by Maxwell ... This provides an analytical proof of the Second Law."[9] This means that according to Boltzmann, since the initial macrostate $[f_0(q,p)]$ can be any macrostate of the form $[f(q,p)]$, the gas that Boltzmann describes can start out in any microstate whatsoever, and still end up within $[f^{MB}(q,p)]$. Thus Boltzmann tried to show that *all* the

---

[9] Boltzmann (1909, I, p. 345) in Uffink (2007, p. 965).

microstates evolve along trajectories that lead them, eventually, to the equilibrium macrostate of the Maxwell–Boltzmann distribution.[10]

It is now easy to see that Boltzmann's argument cannot possibly hold in mechanics, since it counters Liouville's theorem (in a way that we mentioned in Chapter 3). That is, if all the macrostates $[f(q,p)]$, including $[f^{MB}(q,p)]$ itself, are mapped to $[f^{MB}(q,p)]$, the Lebesgue measure of the dynamical blobs is obviously not conserved. Usually the literature on Boltzmann's attempts focuses on two other problems. One is a fault the so-called *Stosszahlansatz*,[11] in the premises of the argument, and the other is the falsehood of the conclusion (disregarding the first fault) known as Loschmidt's objection.[12] This latter objection is very close to one concerning the violation of Liouville's theorem, and we now turn to explain it in some detail.

7.8 Loschmidt's reversibility objection

Consider a gas that expands in a container, as in Figure 1.1. At time t_1 the partition, which confined the gas to the left side of the container, is removed, and – in accordance with the Law of Approach to Equilibrium – at some later time t_3 the gas fills the entire volume of the container. Call the initial macrostate of the gas, which is concentrated in the left-hand side of the container, $[M_0]$, and the final macrostate, in which the gas fills the entire container, $[M_2]$; and call the intermediate macrostate, in which the gas fills, say, about three-quarters of the container, $[M_1]$.

Consider now the *micro*-mechanical description of this evolution, illustrated in Figure 7.4. At t_1 the gas is in microstate a which is in macrostate $[M_0]$. This microstate evolves in accordance with the equations of motion, and at t_1 it arrives at the microstate b, which belongs to macrostate $[M_1]$; the entropy of $[M_1]$ is higher than the entropy of $[M_0]$. Now suppose that at t_2 when the system is in macrostate $[M_1]$ an external agent carries out an instant and simultaneous transformation on all the molecules, in which each of the molecules remains in its position but instantaneously reverses its velocity, so that a molecule that moved to the right now moves to the

---

[10] That this was, indeed, Boltzmann's attempt has been recently argued by Uffink (2007, Sec. 4.2).

[11] The *Stosszahlansatz* is an assumption concerning the number of collisions and is also called the *Molecular Chaos Hypothesis*. It was later proven false (see Ehrenfest and Ehrenfest 1912).

[12] For a recent discussion, see Uffink and Valente (2010).

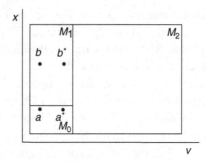

Figure 7.4 Loschmidt's reversibility objection

left, a molecule that moved up now moves down, etc. Such a transformation is called a *Loschmidt reversal*. This reversal results in an instant
change of the microstate of the gas from b to b^*, where b^* is the velocity
reversal of b. As we saw in Chapter 4, in the discussion of velocity
reversals, the consequence of Loschmidt reversal is that as the system
continues to evolve from the reversal time t_2 to time t_3 (where
$t_3 - t_2 = t_2 - t_1$) the molecules will arrive at the microstate a^*, which is
the velocity reversal of a: the velocities of all the molecules in a^* are the
reversals of the velocities they had in the initial microstate a, but their
positions are exactly the same. Thus after the Loschmidt reversal the gas
evolves from the state b^* in which it was spread over three-quarters of the
container to the state a^* in which it is concentrated in the left-hand side of
the container. This seems like a clear violation of the thermodynamic Law
of Approach to Equilibrium.

Loschmidt did not think that such a reversal transformation was practically feasible; indeed, it cannot be carried out even by our most advanced
technologies.[13] However, his argument was that it may happen that a
system will reach microstates such as b^* in the course of its natural evolution. The microstate b^* is in the accessible region of the gas, and so there is
no reason why the gas should not reach such a microstate spontaneously,
and then evolve from higher-entropy to lower-entropy states, in violation
of the Law of Approach to Equilibrium. Since there seems to be no no-go
theorem that would prohibit such an evolution, this is a counter-example
for the claim made by Boltzmann in his equation and *H*-theorem: hence
Boltzmann's argument cannot be valid. It remained to be shown where
exactly was Boltzmann's mistake in his argument (and this mistake was

---

[13] Even in the unique spin echo experiment the reversal is not of velocities, but is carried out
on spins.

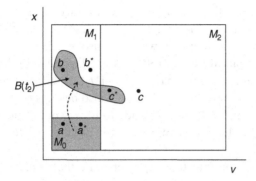

Figure 7.5 Loschmidt reversal

later discovered), but the invalidity of Boltzmann's argument is strictly proven by Loschmidt's counter-example.

It is important to see that the claim that the evolution from b^* to a^* violates the Law of Approach to Equilibrium rests on some assumptions, to which we now turn. On the face of it, Loschmidt's argument showed that to every trajectory that satisfies this law there corresponds a reversed trajectory that violates it. However, the claim that the evolution from b^* to a^* violates the Law of Approach to Equilibrium relies on an assumption concerning the way in which the state space is partitioned into macrostates. In particular, it hinges on the assumption that the microstates a and a^* belong to the same macrostate, namely $[M_0]$, and that the microstates b and b^* belong to the same macrostate, namely $[M_1]$, as in Figure 7.4. This assumption is quite reasonable, but not necessary. In general, in thermodynamic systems, it is reasonable to assume that two microstates that are the velocity reversal of each other belong to the same macrostate. A notable exception to this rule is that wind blowing to the right is macroscopically distinguishable from wind blowing to the left, but this is not the typical sort of macrostate that one has in mind in thermodynamics. A more typical thermodynamically interesting case would be, for instance, a velocity reversal of the air molecules in the room: it is reasonable to assume that such a reversal would be imperceptible.

To complete the picture of Loschmidt's reversibility objection, consider the dynamical blobs involved in it. The dynamical blob $B(t_2)$, which starts out at t_1 in $[M_0]$, definitely contains the microstate b, since b is the end of the trajectory segment which started out in a at t_1; see Figure 7.5. Now since to each microstate within $B(t_2)$ there is a corresponding velocity reversed microstate, we have a velocity-reversed blob $B^*(t_2)$, which returns to the initial macrostate $[M_0]$ in the same time interval $t_2 - t_1$.

Consider now $[M_0]$. The microstate a^* is also (according to our assumption), within the macrostate $[M_0]$, and so the trajectory that starts in a^* at t_1 must end in some point c^* which is also within the blob $B(t_2)$. In Figure 7.5 we depicted the blob $B(t_2)$ as overlapping with both $[M_1]$ and $[M_2]$. Suppose, for example that c^* is in $[M_2]$. A velocity reversal of c^* would take it to some microstate c, which also belongs to $[M_2]$ (according to our assumption), and since a^* leads to c^*, c would lead back to a in violation of the Law of Approach to Equilibrium (as argued by Loschmidt).

Loschmidt's claim, that to every microstate that evolves according to the Law of Approach to Equilibrium, there corresponds a microstate that evolves in an entropy-decreasing way, is a consequence of the velocity-reversal invariance of classical mechanics. This objection reveals that Boltzmann's argument was not a pure derivation from the principles of mechanics but assumed something that has a preference of one time direction over another; and such an assumption cannot be based on pure mechanical considerations.[14]

We can see now in general outline how the Loschmidt objection is related to the violation of Liouville's theorem. In his argument, Boltzmann did not consider all the trajectory segments in the state space. In particular he did not take into account those that reduce entropy. Taking into account all the trajectory segments leads to a violation of Liouville's theorem as well as to Zermelo's objection based on Poincaré's recurrence theorem. We now turn to Zermelo's objection.

7.9 Poincaré's recurrence theorem

Another objection to Boltzmann's argument, put forward by Ernst Zermelo, is based on *Poincaré's recurrence theorem*, which we discussed in Chapter 3 (see Figure 3.9 and accompanying text). As we said, according to this theorem the trajectory of a system starting from any initial state x_0 (except for a set of points of Lebesgue measure zero) will arrive at *some* other time at *any* region around x_0 which is of positive measure. Since the region of the non-equilibrium macrostate $[M_0]$ that contains x_0 as it starts its evolution is of positive Lebesgue measure, the

---

[14] See Uffink and Valente (2010, Appendix) for a discussion on the time-reversal *non-invariance* in Boltzmann's argument.

trajectory that starts out in x_0 will, at some point in time, return to $[M_0]$, in violation of the Law of Approach to Equilibrium.

Notice that, in a sense, Zermelo's objection to the H-theorem is weaker than Loschmidt's objection. Loschmidt's argument means that to every trajectory segment that evolves from a lower-entropy macrostate to a higher-entropy macrostate there corresponds a segment that decreases entropy; whereas the Poincaré recurrence theorem means that every trajectory includes entropy-decreasing segments, without saying that the number of entropy-decreasing segments is equal to or larger than the entropy-increasing segments.

The usual reply to Zermelo's objection is that the recurrence time, after which the system will return close to x_0, is extremely long, so that we are unlikely to experience its effect. But of course, since Boltzmann's H-theorem was meant to be universal, and to underwrite the Law of Approach to Equilibrium in all its generality, this reply cannot save Boltzmann's theorem. It merely emphasizes the need to replace the H-theorem by some probabilistic statement. And so, objections such as those of Loschmidt and Zermelo resulted in Boltzmann's probabilistic turn – that is, in Boltzmann's attempt to prove that the thermodynamic regularities do not hold universally and absolutely, but only probabilistically. Of special importance among these attempts is Boltzmann's combinatorial argument.

7.10 Boltzmann's combinatorial argument

Boltzmann's reply to Loschmidt's reversibility objection was to concede that the mechanical counterpart of the Law of Approach to Equilibrium is not universal but probabilistic: entropy-increasing evolutions are not universal, and the explanation for the fact that we never see anti-thermo-dynamic evolutions is that such evolutions are highly *improbable*. If this claim is cast in terms of the original H-theorem, it would say, roughly, that for *any* initial macrostate $[f_0\ (q,p)]$, *most* of the microstates in this macrostate would evolve, after some suitable time (presumably the relaxation time in which systems typically reach thermodynamic equilibrium), to the macrostate $[f^{MB}\ (q,p)]$.

The inference from low probability to practical impossibility is common. Andrey Nikolaevich Kolmogorov, in his *Foundations of the Theory of Probability*, which laid the foundations for modern probability theory, takes this idea to be one of the fundamental ties between probability and experience: "If $P(A)$ is very small," he says, "one can be practically certain that when conditions C are realized only once, the event A would not occur at all."[15] (Call this the *principle of practical certainty*.[16]) We stress that in the context of statistical mechanics one must be extremely careful in applying the principle of practical certainty, owing to the subtle connection between measure and probability, which we discussed in Chapter 6. In particular, one should be very careful not to infer practical impossibility from a small *measure*, before making sure that this measure is indeed strongly connected to *probability* through our experience with observed relative frequencies. We shall come back to this issue in Chapter 8; now we return to Boltzmann's probabilistic argument.

Boltzmann's argument begins by partitioning the microstates of an ideal gas into macrostates. We have given a detailed description of this partitioning in Section 5.6: briefly, the microstates are first divided into arrangements, where each arrangement includes all the microstates in which the velocities and positions of the particles are distributed over some small intervals, and then the set of arrangements that differ only by permutations of particles are called *distributions*. The distributions are the macrostates in Boltzmann's argument. The next step is to determine the size of macrostates; this is done by counting the number of arrangements in each distribution or macrostate, as is done in the derivation of the MB distribution in Section 5.7. Boltzmann's construction of the arrangements and distributions is such that the number of arrangements in each distribution is finite, and therefore this counting is not problematic. The counting involves combinatorial calculations, and so this argument is sometimes referred to as Boltzmann's *combinatorial argument* or the *permutational argument*.

At this stage of the argument one adds two crucial assumptions, one about *probability* and the other about *entropy*. The first assumption (call it the *probability postulate*) is that all arrangements in the state space of a system are *equiprobable*: that is, if the total number of the arrangements is N then the probability of each arrangement is taken to be $1/N$. This immediately entails that if the number of arrangements that make up a macrostate M is W_M, the probability of M is W_M/N. The second assumption is that the size of a macrostate M, given by the number W_M of arrangements in M, determines the entropy of M, via the expression

---

[15] See Kolmogorov (1933, p. 4).
[16] For the history of this idea, see Hacking (1975). Leibniz, for example, expressed this idea as early as 1668 (p. 146 of Hacking 1975).

$S_M = k \log W_M$. We already know (see Section 5.7) that entropy thus defined is strongly connected to thermodynamic entropy via the Maxwell–Boltzmann distribution, which characterizes the maximal W_M. We can now write the expression for entropy as $k \log (W_M/N) + k\log N$, where N is the total number of arrangements of the system. And since $k \log N$ is constant for this system, it cancels out in calculations of entropy differences (recall from Chapter 2 that it is entropy differences that matter in thermodynamics). And therefore we have for all practical purposes $S_M = k \log (W_M/N)$, in which case it appears that the entropy of M is given by M's probability; call this the *entropy-as-probability idea*.

This looks like a sound argument. Once we accept what entropy is in statistical mechanics and the combinatorial idea that all arrangements are equiprobable, the next steps in the argument follow logically. And moreover the argument is very tempting since it seems to entail *directly* the conclusion that a system is more likely to be in macrostates that have higher entropies; and this sounds very close to a mechanical underwriting of a probabilistic version of the Law of Approach to Equilibrium. Indeed, this idea of entropy-as-probability was very influential in statistical mechanics and has remained prevalent in the literature until today.

Nevertheless, Boltzmann's combinatorial argument is highly problematic. To start with, it is not clear what the justification is for the probability postulate. When the number of arrangements is discrete it may *prima facie* seem an almost analytical truth that the probability of each of the N possible arrangements is $1/N$ (but in Chapter 8 we argue that this idea is unjustified). And if so, the probability of a macrostate in the state space should be given by the total number of arrangements that make up this macrostate divided by the total number of arrangements, that is W_M/N. But once we take into account the dynamics of the system, this reasoning is no longer compelling. Indeed, it seems to be an equally analytical truth that, since the evolution of thermodynamic systems is governed by their mechanical dynamics, and since this dynamics can give rise to a variety of evolutions of the dynamical blob, the transition probability from a given macrostate to any other macrostate at any given time can be almost anything we want. Boltzmann's combinatorial argument makes the dynamics of the system irrelevant for describing its evolution. It rules out the possibility that *a large measure* of the trajectories that start out in non-equilibrium macrostates may not evolve to equilibrium, owing to the details of the dynamics. But nothing in mechanics prevents this case; in particular, Liouville's theorem is satisfied by a dynamics in which all the microstates of a system remain in their initial macrostates for all time (compare the discussion in Section 6.4). By contrast, in our above formulation of the mechanical counterparts of the Law of Approach to Equilibrium and the Second Law, the probabilistic claims are directly linked to the dynamics, and this fact makes these laws non-trivial and in need of proof. Moreover, since Boltzmann's probability postulate is non-dynamical, it is not even clear what the *meaning* is of the statement that the probability of each arrangement is $1/N$. We discuss this particular question in Chapter 8.[17]

This criticism of Boltzmann's combinatorial approach has an implication concerning the derivation of the Maxwell–Boltzmann distribution described in Section 5.7. Recall that we stressed that this derivation refers to the size of macrostates as measured by the number of arrangements, and not to the probability of macrostates. The arguments above show that this point is crucial to the understanding of probability in statistical mechanics. That is, the combinatorial argument for deriving the MB macrostate shows (i) that the MB macrostate characterizes the thermodynamic notion of equilibrium (by deriving from it certain successful predictions); and (ii) that the MB macrostate is the largest macrostate, by counting the number of arrangements compatible with it. This property may reasonably be associated with high entropy. However, the combinatorial argument fails to prove that the MB macrostate is the most probable one in the sense that the system is highly likely to evolve to it from its initial macrostate.

Finally, it turns out that Boltzmann's probabilistic postulate is empirically inadequate; it is refuted by the most trivial experience with thermodynamic systems. The reason is that it gives the wrong description of the well-known phenomenon of the approach to equilibrium, such as that described in Figure 1.1: it implies that the gas is likely to evolve *directly* from the state in which it fills half of the container to the state where it fills the entire container, not going through the intermediate stages where it fills, say, three-quaters of the container. Here is why. According to Boltzmann's argument the probability of a macrostate is given by its size, which is time-independent. Suppose now that a system starts in a non-equilibrium macrostate $[M_0]$, and ends, after some time, in an equilibrium macrostate, $[M_2]$. In general, on its way from $[M_0]$ to $[M_2]$ the system passes through other non-equilibrium macrostates, such as $[M_1]$ (see for example Figure 1.1 and Figure 7.3). Now, according to the probability postulate, since the

---

[17] The Principle of Indifference is sometimes cited in this context; see Chapter 8.

number of arrangements in $[M_2]$ is larger than the number of arrangements in $[M_1]$, the probability of $[M_2]$ is higher than the probability of $[M_1]$. This means that at t_1, when the system just starts out in $[M_0]$, it is already the case that the transition probability to $[M_2]$ is larger than the transition probability to $[M_1]$: $P([M_2]) > P([M_1])$, and this means that the system is more likely to evolve directly to $[M_2]$ than it is to evolve directly to $[M_1]$. Nothing in Boltzmann's combinatorial argument suggests that the system is likely to pass through $[M_1]$ on its way to $[M_2]$. This goes against our experience, in which systems evolve to equilibrium gradually, via non-equilibrium macrostates. Moreover, this may also go against the limitations posed by the special theory of relativity, since it entails that a system can jump from one set of microstates to another, while there is no guarantee that the microstates in these sets are time-like separated.

For these reasons it seems to us that Boltzmann's earlier approach, namely the approach that underlies his equation and the *H*-theorem, which is fundamentally based on the dynamics of the system, is far better than his later approach. It seems to us that in the Boltzmann equation Boltzmann had in mind something like a dynamical evolution of macrostates, which roughly underlies our approach in the Probability Rule. There have been attempts to replace Boltzmann's original argument with probabilistic ones that do not suffer from the problems that beset that argument. Of course, such proofs cannot be general and universal, since the Law of Approach to Equilibrium has counter-examples in the form of Maxwell's Demon. Nevertheless, as we said above, since we are surrounded by systems that seem to satisfy this law, it is reasonable to expect that one can prove probabilistic versions of Boltzmann's equation and *H*-theorem, for interesting special cases. Moreover, since the laws of thermodynamics are time-asymmetric, it is clear that in any such proof one must make a time-asymmetric assumption, and the question is whether such an assumption is acceptable. One of the most important attempts at such a proof is by Oscar E. Lanford, and we now turn to discuss it.

7.11 Back to Boltzmann's equation: Lanford's theorem

Lanford (1975, 1976, 1981)[18] proved a theorem of statistical mechanics that attempts to underwrite a probabilistic version of the Law of

---

[18] The original theorems are in Lanford (1975, 1976, 1981); For a recent comprehensive discussion, see Uffink and Valente (2010).

Approach to Equilibrium. Lanford worked in the direction that Boltzmann started, in the Boltzmann equation and the *H*-theorem, and can be understood as proving them for special cases. In this section we will cast Lanford's theorem in our terms of dynamical blobs and macrostates, and describe it in general terms.

Lanford's theorem holds for an ideal gas in which the particles are treated as hard spheres, in the highly idealized conditions of the so-called *Boltzmann–Grad limit* (that is, the gas is taken to consist of infinitely many particles confined to a limited volume, where the density is kept constant by taking the diameter of the particles as approaching zero, though not becoming zero, so that one can still treat them as hard spheres). The hard spheres model determines the dynamics of the system: that is – in our terms – it determines the evolution of the dynamical blob; and the Boltzmann–Grad limit characterizes the initial macrostate $[M_L]$ of the system. The theorem proves that a system that starts in $[M_L]$ approaches equilibrium within a time interval Δt_L,[19] in the case where the initial macrostate $[M_L]$ is initially not too far from the relevant equilibrium macrostate $[M_{eq}]$. Specifically, the spatial distribution is not too far from the uniform distribution, and the velocity distribution is not too far from Maxwell's distribution.[20] Lanford's theorem, cast in our terms, proves that the evolution of the dynamical blob $B(t)$ of the system, which starts in the initial macrostate $[M_L]$, is such that among the microstates in the initial macrostate $[M_L]$, a Lebesgue measure 1 of microstates will arrive at the equilibrium macrostate $[M_{eq}]$ within the short time interval Δt_L. There are microstates in $[M_L]$ that will not approach equilibrium within Δt_L, but their set has a Lebesgue measure 0. Here, Lanford implicitly applies the *principle of practical certainty*, which justifies the conclusion that such non-thermodynamic microstates will not be experienced.

Despite all these assumptions and limitations, Lanford's theorem is one of the most important achievements in this field, and for this reason it is important to understand what exactly it proves. One point which calls for discussion here is an issue we began to discuss in Chapter 6 (and will continue to discuss in Chapter 8), namely the issue of the connection between measure and probability.

In the usual way of thinking about Lanford's theorem, the claim that a set of Lebesgue measure 1 of the microstates in the initial macrostate $[M_L]$ evolves to equilibrium within the time interval Δt_L is understood as

---

[19] For discussion of this time interval, see Uffink and Valente (2010) pp. 156–8.
[20] See Uffink and Valente (2010) p. 155.

entailing (or *meaning*) that the *probability* that a microstate that is randomly sampled out of $[M_L]$ will approach equilibrium within Δt_L is 1. This understanding is mistaken, since it is in general erroneous to deduce probability from measure.[21] We expand on this point in Chapter 8; here we discuss it only in the special case of Lanford's theorem.

As we said in Chapter 6, the measure of probability, i.e. the measure μ in the Probability Rule, ought to be chosen according to the observed relative frequencies. Hence, arguments about measure do not entail probabilities, but are *inferred from* probabilities. In Lanford's theorem, the transition probability from $[M_L]$ to $[M_{eq}]$ is given by the measure μ of the overlap of the dynamical blob that starts in $[M_L]$ at t_0 and ends in $[M_{eq}]$ at $t \leq t_0 + \Delta t_L$, with $[M_{eq}]$. This overlap should be measured by the most convenient among the measures that reproduce the observed relative frequencies of evolutions from $[M_L]$ at t_0 to $[M_{eq}]$ at $t \leq t_0 + \Delta t_L$. If, as a *contingent* fact, it so happens that the observed relative frequency of the evolutions from $[M_L]$ at t_0 to $[M_{eq}]$ at $t \leq t_0 + \Delta t_L$ is (approximately) 1, *and if* the most convenient measure of the overlap between the blob and macrostates that reproduces this observation happens to be the Lebesgue measure (or some measure that is absolutely continuous with it), which is conserved under the dynamics, *then* one may legitimately describe Lanford's theorem as describing the behavior of Lebesgue measure 1 of the microstates in $[M_L]$. In other words, Lanford's theorem may be understood as a theorem about the probability of evolving to equilibrium only if the measure in the Probability Rule turns out to be the Lebesgue measure (or any other measure absolutely continuous with it). Thus, fundamentally, the measure-1 statement does not *entail* a probability 1 statement concerning the behavior of microstates, but is rather a *consequence* of a probability 1 statement based on observed relative frequencies.

Of course, the way we described things in the above paragraph is not how the proof of Lanford's theorem looks. Formally, Lanford (and others who developed his idea) do not rely explicitly on observed relative frequencies in their arguments. However, the very choice of the Lebesgue measure in this argument has *no other justification*, except the empirical adequacy of this choice. One may read Lanford's theorem as a conjecture, which would be either confirmed or disconfirmed by experiments with systems that are reasonably approximated by Lanford's model of hard spheres in the Boltzmann–Grad limit. Alternatively, given that, in our

---

[21] Recall that we encountered a similar claim in our discussion of the ergodic theorem in Section 3.7.

experience, there is quite a high probability that thermodynamic systems will approach equilibrium within the time interval described by Lanford (provided that they are initially not too far from equilibrium) one may conclude that Lanford's model of hard spheres in the Boltzmann–Grad limit is a good approximation of the observed thermodynamic systems.

Finally, a very important aspect of Lanford's theorem is its *time asymmetry*. Since mechanics is time-symmetrical (in the sense described in Chapter 4), it is clear that the theorem is based on some non-mechanical assumption, which puts time asymmetry in, by hand. A careful analysis of Lanford's theorem reveals the culprit.[22] However, the important lesson is that any attempt at underwriting the Law of Approach to Equilibrium by mechanics must be based on a time-asymmetric assumption that does not result from the principles of mechanics. The question is which such assumption (if any[23]) is acceptable. This issue is central for underwriting thermodynamics by mechanics, and we shall deal with it in Chapter 10.

7.12 Conclusion

The fact that in Lanford's theorem, and in all attempts to underwrite the thermodynamic regularities by mechanics, the direction of time has to be inserted by hand, is no accident, and cannot be avoided, owing to the time symmetry of mechanics. By way of summing up this chapter, we wish to emphasize some important aspects of this issue. These are related to our discussion of time-reversal invariance in Chapter 4.

In Chapter 4 we introduced the notion of time-reversal invariance in classical mechanics. We said that this invariance can be understood in terms of the fact that an observer cannot distinguish between the two possible directions of time: the one in which the possible evolutions are (I)–(IV) and the one in which the possible evolutions are (I*)–(IV*). For a figurative illustration we said that we experience the first direction, and an observer called Tami experiences the time-slices in the reversed order and with the opposite direction of time built into each of the time-slices. We emphasized that we and Tami agree on the spatial content of the time-slices and on their relation of betweenness; the only difference between us is the direction of time. Mechanics is *time-reversal invariant* in that the laws of mechanics hold equally for us and for Tami, and there is no

---

[22] See Uffink and Valente (2010).
[23] Huw Price takes this issue to be central in what he calls the "view from nowhen."

mechanical experiment that each of us can carry out that will reveal one time direction as distinct from the other. The *only* thing we can say about Tami's universe is that its time flows in a direction opposite to ours.

However, when we add the thermodynamic regularities to this picture, things change: clearly, if a gas expands in our time direction, Tami will experience this gas as contracting. In brief, Tami will summarize her experience in a theory that will be *anti-thermodynamics*.

However, given that we and Tami agree on the laws of mechanics, we ought to agree on every theorem of mechanics, including every probabilistic theorem of mechanics. And if thermodynamics can be underwritten by probabilistic theorems of mechanics, then thermodynamics has to hold in both directions of time! Replacing absolute theorems (such as Boltzmann's *H*-theorem) with probabilistic theorems does not solve the problem, since, if mechanics tells us that systems are highly likely to obey the Law of Approach to Equilibrium, then this should hold for Tami as well. Given that it is a theorem of mechanics, the mechanical probabilistic counterpart of the Law of Approach to Equilibrium should be *either correct both for us and for Tami, or false for both*.

If the mechanical probabilistic counterpart of the Law of Approach to Equilibrium is correct both for us and for Tami, then Tami should predict that entropy must increase in the thermodynamic systems that she observes, but given that entropy consistently increases in our time direction, Tami will fail in her predictions time and again. If she adheres to the theorem despite these failures, she will conclude that her actual trajectory is *atypical*: despite the high likelihood of entropy increase, she happens to experience the unlikely cases, over and over again.

But such a consistent failure will happen only if Tami adopts the measure μ of probability and the measure ν of entropy that fit *our* experience. If she replaces them with measures that reflect the relative frequencies that *she* observes and the degree to which *she* can exploit energy, her theory will be good and useful – albeit anti-thermodynamic relative to our measures.

Our present discussion is not about the metaphysics of a possible universe with a reversed time direction. The reason we take the trouble to analyze the case of Tami's universe is in order to learn about our world, with our direction of time. And here there are important lessons to be learned.

First, a *regular* anti-thermodynamic behavior is consistent with mechanics. And so there is no *a priori* reason to rule out the possibility that it will happen in our world, starting five minutes from now. There is not, and there cannot be, a general theorem to the effect that anti-thermodynamic behavior is unlikely. At best, we can expect such theorems in very special circumstances, and in which the direction of time is inserted by hand, as in Lanford's theorem.

The second lesson is that the measure μ of probability and the measure v of entropy are chosen so as to reflect our actual empirical experience. There is nothing *a priori* about them, and no measure has an *a priori* advantage over other measures in this context.

The measure μ of probability is chosen on the basis of observed relative frequencies, and the measure v, by which we determine the entropy of macrostates in statistical mechanics, is chosen by comparing the statistical mechanical entropy to thermodynamic entropy, which expresses observed degrees of energy exploitability. By choosing the measure v of entropy, we achieve two aims: one is that, conceptually, we adhere to the idea that entropy is associated with the degree to which we can *control a microstate*, and this degree is, in turn, associated with the size of the macrostates. The other aim is that the magnitude we call entropy in statistical mechanics ought to correspond to the thermodynamic magnitude, which quantifies the degree of *exploitability of energy*, as we said in Chapter 2.

Now, the very association of the magnitude of entropy with the degree of exploitability of energy goes via the Law of Approach to Equilibrium and the Second Law of Thermodynamics, as formulated by Kelvin and Clausius. This means that the definition of entropy is appropriate in a world in which these regularities hold. However, as we emphasized in Chapter 2, the thermodynamic regularities are nothing but empirical generalizations. They express our generalized experience. There is nothing *a priori* about them (despite expressions to the contrary by the likes of Einstein and Eddington; see Chapters 1 and 13). A world in which the thermodynamic laws are not satisfied is conceivable; it does not contain a contradiction. Moreover, a world that violates the thermodynamic generalizations does not violate the laws of mechanics, nor the principles of statistical mechanics described so far in this book. We prove this claim in Chapter 13. Now, if the very definition of entropy in thermodynamics is based on the validity of the thermodynamic laws, then in a world that violates these laws the thermodynamic notion of entropy will no longer correspond to exploitability of energy. In other words, in a Demonic world, in which Maxwellian Demons exist, the very definition of entropy will have to change: the size of a macrostate will no longer be inversely correlated with the degree of exploitability of energy. But then, in such a world, the whole conceptual structure of thermodynamics, as presented in Chapter 2, would not be suitable for describing nature. In such a world we would still have mechanics, and would still have statistical mechanics; but we would not have thermodynamics as we know it, with all its laws and concepts.

8

Typicality

8.1 Introduction

In the previous chapters we outlined the way in which we think probability should be understood in statistical mechanics, namely in terms of the Probability Rule (Chapter 6). But our view is different from some mainstream views about probability in statistical mechanics, particularly from a view which has recently come to be known as the *typicality approach*.[1] We distinguish between two kinds of typicality considerations and explain why only one of them seems to us acceptable in a physical theory.

An essential part of our argument is the distinction between a probability measure and physical probability. This idea is expressed by Itamar Pitowsky:

> Consider a finite but large collection of marbles. When one says that a vast majority of the marbles are white, one usually means that all the marbles except possibly very few are white. And when one says that half the marbles are white, one makes a statement about counting, and not about the probability of drawing a white marble from the collection. The question is whether non-probabilistic notions such as *vast majority* or *half* can make sense, and preserve their meaning when extended to the realm of the continuum. In particular, when the elements of the collection are the possible initial conditions of a large physical system.[2]

We will address and explain this distinction between measure and physical probability as we proceed. But first, let us introduce the typicality approach.

---

[1] For a presentation of this approach by one of its proponents, see Goldstein (2012).
[2] Pitowsky (2012).

8.2 The explanatory arrow in statistical mechanics

The aim of statistical mechanics is to predict the behavior of systems when given only incomplete information about their actual microstate, in terms of their macrostate; and the conceptual tool with which statistical mechanics pursues this aim is probability. Without going into the various interpretations of the notion of probability, we take the following as our guidelines in dealing with probability in physics and, particularly, in statistical mechanics. The probability that a system will be in a certain macrostate $[M_1]$ in the future, given the system's initial macrostate $[M_0]$ and the system's dynamics, is closely related to the relative frequency with which such systems have been observed to evolve from $[M_0]$ to $[M_1]$ in the past. The statement that the transition probability from $[M_0]$ to $[M_1]$ is p, is associated with both the claim that p reflects past relative frequencies of such transformation, and the expectation that these relative frequencies will be approximately repeated in the future. This is essentially our requirement of *testability* mentioned in Chapter 6. We shall call a probability measure that satisfies this requirement *physical probability*. All this is very general, but it is enough to start our present discussion.

This idea motivates our interpretation of the role and status of probability in statistical mechanics in terms of the Probability Rule. In particular it motivates the way we think one should determine the measure μ in this rule. As we explained in Chapter 6, probability comes into play in statistical mechanics owing to the relations between macrostates and dynamical blobs, and the central probabilistic statement of statistical mechanics is the Probability Rule. For a system that starts in macrostate $[M_0]$ at time t_1, the probability that it will end in macrostate $[M_1]$ at time t_2 is given by the measure μ of the overlap of the state-space region of the dynamical blob $B(t_2)$ with the state-space region $[M_1]$. (See, for example, Figure 6.2). In our approach, the measure μ gives the probability of transition between macrostates, *if* it is chosen and updated so as to fit the relative frequencies, observed in the past, between the macrostates $[M_0]$ and $[M_1]$ during time intervals equal to t_2-t_1. Only in this case μ is an empirically significant basis for predictions.

It is crucial to note already at this stage that the Probability Rule is *not* derived from the principles of mechanics, since the measure μ is a generalization drawn from experience beyond that on which the principles of mechanics are based. The fact that the Probability Rule cannot be derived from mechanics alone means that statistical mechanics involves two distinct inductive steps: first, it is based on the inductive generalizations underlying the laws of mechanics such as $F = ma$. But it is also based on *further* inductive generalizations, *not* derivable from the first, namely generalizations concerning the relative frequencies of macrostates, that lead to the choice of the measure μ in the Probability Rule. By contrast, in the typicality approach the idea is that the probabilistic counterparts of the laws of thermodynamics can be derived from the principles of mechanics, together with some *a priori* considerations. In this, it seems that the typicality approach follows the tradition that originated with Boltzmann's combinatorial approach (see Section 7.10).[3]

This point can be understood in terms of the explanatory arrow. In our approach, experience explains the choice of the measure μ in the Probability Rule.[4] In the typicality approach, the direction of the explanatory arrow is reversed: the probabilistic statements are taken to explain our experience.

8.3 Typicality

The typicality approach consists of three statements together with the direction of the explanation (as described above). The three statements are the following:

(i) The set of initial conditions compatible with the initial macrostate $[M_0]$ is divided into two subsets, call them T_1 and T_2, so that all the microstates in T_1 but not in T_2 evolve in a certain way, say, approach the equilibrium macrostate within the relaxation time.

This is a *contingent fact* about the dynamics. There are theorems in classical statistical mechanics that demonstrate that special cases of (i) hold under some conditions. For example, Lanford's theorem (described in Section 7.11 and discussed below) shows that (i) holds for special initial macrostates and special dynamical conditions. We take (i) to be uncontroversial.

(ii) There is a measure over the state space (call it L), such that $L(T_1)$ is close to 1 and $L(T_2)$ is close to 0. In this sense, *most* of the initial microstates that are compatible with the initial macrostate $[M_0]$ are in T_1, and are called *typical*. This statement is not committed to a specific probability measure L: "Insofar as typicality is concerned, the

---

[3] The same idea is expressed in Goldstein (2012, Sec. 3), although with a focus on Boltzmann's Equation and Lanford's approach discussed here is Section 7.11 and below.
[4] See also on this point our (2011a).

detailed probability of a set is not relevant; all that matters is which sets have very large measure and which very small."[5]

Statements (i) and (ii) together entail that typical microstates in the initial macrostate evolve to equilibrium.

(iii) In a given experiment, the actual initial microstate of the system is highly likely to belong to the *majority* set T_1.

Statements (i), (ii), and (iii) together entail that in a given experiment the system is highly likely to evolve to equilibrium.

The notion of *most* in statements (ii) and (iii) above is about a measure over the phase space. There are many ways to determine the size of subsets of a continuous set of points, and the question is on what grounds one can justify the choice of measure, or the choice of some class of measures, in order to determine the size of a set of points in a given case.

It is this question that distinguishes between the two readings of the typicality approach. The first (which we call *a posteriori*) is the one we have advocated in the choice of the measure μ for the Probability Rule which, as we said in Chapter 6, is based solely on experience (and some pragmatic considerations). If it turns out that the μ measure in the Probability Rule is a measure that is conserved under the classical dynamics (for example the Lebesgue measure), then one can follow the time sequence of the dynamical blobs in reverse, as it were, to $[M_0]$, and say that the probability of a set of points in $[M_0]$ evolving to a given macrostate is proportional to this set's Lebesgue measure. On this reading, the Lebesgue measure is (merely) a convenient measure that fits the observed relative frequency of macrostates. And it is precisely this empirical and pragmatic justification for the use of the Lebesgue measure which makes this reading of statement (ii) acceptable.

It is crucial to see that in this reading of (ii), the justification of the choice of the measure over the initial macrostate is *grounded* in experience, is a *generalization* of experience, and therefore it cannot be taken to *explain* experience in a non-circular way.

On the other hand, it may turn out that the right measure to use in the role of μ in the Probability Rule is not conserved under the dynamics. In this case, the transition probability from $[M_0]$ to some macrostate $[M_i]$ will not be equal to the μ measure of the set of points in $[M_0]$ that evolve to $[M_i]$. In other words, in this case the μ measure (which is the right probability measure to use) of the set of points in $[M_0]$ will not yield the right predictions. Note in passing, as we stressed several times, that in

[5] See Goldstein (2012, Sec. 5).

both cases the probability of transition to a macrostate $[M_i]$ as given by the Probability Rule is in general unrelated to the μ measure of $[M_i]$.

The *a posteriori* reading of (ii) (when applicable) seems to us acceptable provided that one remembers the direction of the explanatory arrow. It is the other reading of (ii) with its reversed direction of the explanatory arrow (which we call the *a priori* reading) that we believe is indefensible. On this reading, the choice of the measure L in (ii) is *dictated* by the fact that L has some preferred dynamical status in the theory, sometimes together with *a priori* considerations. For this reason, it is argued, our experience is explained by the L measure and not *vice versa*. As we said before, an example of this reading is Boltzmann's combinatorial approach, and in particular Boltzmann's probabilistic postulate according to which all the arrangements are equally probable (see Section 7.10). Another example is the ergodic approach as expressed by Reichenbach (see Section 3.7).

From now on, our discussion of the typicality approach focuses on this *a priori* reading. We now turn to criticize the *a priori* reading of (ii) and then the transition from (ii) to (iii) in this reading.

8.4 Are there natural measures?

An argument sometimes given for preferring the Lebesgue measure as "natural" on the basis of the classical dynamics is the invariance of the Lebesgue measure under the dynamics as expressed by Liouville's theorem. Of course, given Liouville's theorem, the Lebesgue measure has very attractive properties, such as simplicity and elegance, and so it is tempting to use it in as many contexts as possible. However, it is unclear why Liouville's theorem should be relevant at all to the issue at stake in the *a priori* approach. We saw that Liouville's theorem puts a constraint on the evolution of the blob, but it plays no role in determining the relative size of the regions of overlap between the blob and the macrostates. It is only in the *a posteriori* approach that the conservation of the measure (associated with Liouville's theorem) can be relevant at all for the distribution of probability over initial conditions.

A similar argument for preferring the Lebesgue measure as "natural" is sometimes put forward in the case of ergodic dynamics. Obviously, the ergodic theorem gives a preferred status to the Lebesgue measure (or to any measure absolutely continuous with it) for two reasons. First, the theorem shows that if the dynamics is ergodic, then a set of Lebesgue measure 1 of the microstates of the system lies on trajectories along which the infinite relative frequency of any macrostate is equal to the Lebesgue measure of that macrostate. Second, if the dynamics is ergodic, then only measures absolutely continuous with the Lebesgue measure are conserved under the dynamics. However, the preferred status of the Lebesgue measure here is irrelevant for the *a priori* reading of the typicality approach, essentially since the ergodic theorem is about measure and not about physical probability. The reason is that the ergodic theorem yields no predictions concerning finite times, and therefore strictly speaking the theorem is not empirically testable. For example, it is extremely difficult to distinguish empirically between an ergodic system and a system with other forms of complex dynamics.[6] Notice that although the ergodic theorem is usually understood in terms of physical probability, it is a theorem about measure; and since it yields no definite predictions of relative frequencies in finite times, it does not satisfy the above requirement for interpreting a measure as physical probability.

A third argument sometimes given for taking the Lebesgue measure as the natural measure in statistical mechanics is that the Lebesgue measure of a macrostate corresponds to the thermodynamic *entropy* of that macrostate. However, this correspondence is true only if the Law of Approach to Equilibrium in its probabilistic version is true (see Section 7.12), and only if the v-measure of entropy is indeed the Lebesgue measure (see Section 7.2). However, since the choice of the measure of entropy v is rooted in experience, this argument is either *a posteriori* or begs the question.

8.5 Typical initial conditions

Statement (iii), as given above, seems to be expressing the brute fact, without further reasoning, that in every experimental set up satisfying some known constraints, the microstate is highly likely to belong to the majority set T_1; the microstate in the minority subset T_2 has only a slight probability of being realized. But since T_2 is not empty, this fact calls for a justification. Statement (ii) is meant to *explain* the observed thermodynamic behavior of physical systems by entailing statement (iii). Statement (ii) becomes empirically significant provided it implies a statement about probability.

We now turn to criticizing the transition from statement (ii) to statement (iii) in the *a priori* reading.

---

[6] See Earman and Redei (1996). For example, it is hard to distinguish empirically between ergodic dynamics and the disordered regions in the state space of a system with KAM dynamics (a result obtained by Andrey Kolmoyorov, Vladin Arnold and Jürgen Moser; see Walker and Ford 1969).

The derivation of (iii) from (ii) is based on the fact that formally, in the *mathematical* theory of probability summarized in Kolmogorov's axioms, probability is a kind of measure. If indeed any measure that satisfies the appropriate conditions could be taken as physical probability, then (ii) would have led to (iii). But the notion of probability *in physics* has a meaning above and beyond this formal requirement. Measure should be understood as *physical* probability only if it has empirical significance. And whereas the measure μ in the Probability Rule (and the *a posteriori* reading of typicality) satisfies this requirement, the preference for the *L*-measures in the *a priori* reading does not, and therefore it cannot be interpreted as physical probability.

In both readings of the typicality approach, (iii) is taken to be justified by (ii). The idea is that microstates in low-*measure* sets have a low probability of being realized. In the *a posteriori* reading this idea is true *by construction*, since the measure is chosen on the basis of experience. In other words, we start with physical probability that reflects actual relative frequencies, and then we choose a measure that conveniently describes our experience; for example we take a measure relative to which a uniform distribution gives a large value to sets of high probability. We could choose other measures to describe the same situation in a less elegant or convenient way. In this view, what causally explains the transition from (ii) to (iii) is not the measure but the physical probability.

By contrast, in the *a priori* reading the distinction between measure and physical probability becomes crucial: the justification of (iii) contains an assumption that sets have high physical probability *because* they have large measure. Here is Detlef Dürr's way of putting the matter:

> What is typicality? It is a notion for defining the smallness of sets of (mathematically inevitable) exceptions and thus permitting the formulation of law of large numbers type statements. Smallness is usually defined in terms of a measure. What determines the measure? In physics, the physical theory. Typicality is defined by a measure on the set of "initial conditions" (eventually by the initial conditions of the universe), determined, or at least strongly suggested by the physical law. Are typical events most likely to happen? No, they happen *because* they are typical. But are there also atypical events? Yes. They do not happen, because they are unlikely? No, *because* they are atypical. But in principle they could happen? Yes. So why don't they happen then? Because they are not typical.[7]

---

[7] Dürr (2001, p. 131), our emphasis.

But since there are infinitely many ways to put a measure over the state space, all of which can be said to describe the same physical process, it follows that the measure cannot be taken to explain the process. It is in this reversal of the direction of explanation that the *a priori* reading of typicality puts the cart before the horse. We believe that this is what Pitowsky meant in the quotation above.

Prima facie, one could try to address this difficulty by postulating a probability distribution (say, a uniform distribution relative to the *L*-measure[8]) over the initial conditions. Here, the probability distribution is meant to have a physical content, namely to describe relative frequencies. But relative frequencies of what? Usually, the idea is that the probability distribution describes a random sampling of microstates out of the initial macrostate of some subsystem of the universe (say, a gas in a box). This means that the probability of randomly sampling a microstate from a given region in the initial macrostate is equal to the *L*-measure of that region. The physical significance of this idea is that there is some random state generator external to the subsystem, which prepares the subsystem in its initial microstate. But if this random state generator is itself a mechanical system, then its randomness can come only from some other external random state generator, and so on until the beginning of the universe. But then it turns out that the first random state generator of the microstate of the universe is external to the universe. Therefore, it cannot be physical.[9]

8.6 Measure-1 theorems and typicality

Using the ideas developed above, let us assess the significance of Lanford's theorem, mentioned in Section 7.11. This theorem is an example of measure-1 theorems, and the way we analyze it below is an example of the way in which the significance of measure-1 theorems in general should be understood.

As we said in Section 7.11, on the basis of the classical equations of motion, Lanford proved that if a system starts at time t_0 in the macrostate $[M_L]$, and its dynamical blob evolves according to Hamiltonian H_L then the

---

[8] This is Albert's (2000, Ch. 4) Statistical Postulate; see Section 6.9.
[9] In quantum mechanics (in its probabilistic versions, such as the Ghirardi, Rimini, and Weber (GRW) theory), we have a probability distribution built into the dynamical equations of motion. But still to get the relative frequencies of the events we experience with high (quantum mechanical) probability we need a *suitable* initial wavefunction of the universe.

Lebesgue measure of the overlap between the blob and the macrostate $[M_{eq}]$ of equilibrium is 1. Of course, since the Lebesgue measure is conserved under the dynamics, one may employ here the *a posteriori* reading of the typicality approach; it is only the *a priori* reading that is problematic. The right way to understand Lanford's theorem is, therefore, the following. The theorem is about measure. In order to see whether or not this measure corresponds to physical probabilities, one has to carry out observations on systems that start out in $[M_L]$ and evolve according to H_L, and see the relative frequencies of their macroscopic evolution to $[M_{eq}]$. If these relative frequencies turn out to be such that the Lebesgue measure fits them, then we can take this measure to describe probabilities. But it is crucial to notice that the justification for using this measure as expressing physical probabilities rests on the empirical generalization from experience concerning relative frequencies, and not on the theorem. In the terms used above, the general structure of Lanford's theorem is that it shows that: (i) a certain subset of microstates T_1 in the initial macrostate $[M_L]$ share the significant property of approach to equilibrium within the relaxation time; and that (ii) the subset T_1 has Lebesgue *measure* 1. But it does not show that this measure corresponds to the physical probability that can give rise to the statement (iii) that a given system of interest is highly likely to have its microstate in T_1.

In the literature, typicality results are often formulated in terms of Laws of Large Numbers of probability theory.[10] However, such formulations are not immune to our criticism, since they too require a choice of measure over the sequences. This choice needs to be justified for exactly the same reasons spelled out above.

Typicality considerations of the kind described in this chapter are abundant, and suffer the same kind of conceptual problems. An interesting example that we only mention in passing is Einstein's[11] account of the Brownian motion, as developed by Wiener.[12] Wiener has proved that the so-called Wiener measure of trajectories in the state space of a Brownian particle, which are continuous but nowhere differentiable, is 1. The explanation of the actual behavior of Brownian particles is based on the assumption that their actual trajectories belong to this measure-1 set, and Avogadro's number is derived from this assumption.

Simple measures, which have special dynamical properties, and for which we have measure-1 theorems, are very important in physics. But

---

[10] See, for example, Dürr, Goldstein and Zanghi (1992), Callender (2007), Goldstein (2012).
[11] Einstein (1905).
[12] See Pitowsky (1992).

their importance lies only in the fact (to the extent that it is a fact) that they fit observed relative frequencies. Their elegance and their dynamical properties have nothing to do with the fact that they reflect probabilities. The empirical success of some of our best theories, including statistical mechanics, makes it good heuristics to start out by testing such measures. "But," to quote van Fraassen, "as we know, this method always rests on assumptions which may or may not fit the physical situation in reality. Hence it cannot lead to *a priori* predictions. Success, when achieved, must be attributed to the good fortune that nature fits and continues to fit the general model with which the solution begins."[13]

8.7 Conclusion

What is the significance of measure-1 theorems in statistical mechanics? Suppose that we are given some measure-1 theorem, say Lanford's theorem, and suppose that we are given a system that satisfies all the conditions presupposed by the theorem. Can we predict that the system will evolve to equilibrium in the designated time with *probability* 1? The answer is *No*. The reason is that in order to interpret the *measure* as denoting *physical probability* one has to add further knowledge about the system which is not given in the theorem nor anywhere else in mechanics: it is the knowledge that enables us to infer physical probability from measure. This knowledge is about the relative frequencies of macrostates. These relative frequencies lead us to the choice of the measure μ in the Probability Rule. It may turn out that the measure μ is the measure that appears in the theorem, but nothing in the principles of mechanics guarantees that this is always going to be the case. In order to arrive at probabilistic statements concerning thermodynamic behavior we must carry out two inductive steps: First we need induction in order to arrive at the principles of mechanics, and then we need a further and independent induction over the relative frequencies of macrostates in order to arrive at the Probability Rule. The aim that seems to motivates the typicality approach, to derive the statistical mechanical probabilities from mechanics together with probability theory alone, without recourse to further inductive generalizations from experience, is hopeless.[14]

---

[13] van Fraassen (1989, p. 316).
[14] The case in which probability already appears in the laws of mechanics, is discussed in Appendix B.1.

9

Measurement

9.1 Introduction

Up to now, our descriptions of thermodynamic evolution have started with the somewhat bold assumption of an initial macrostate $[M_0]$ of a thermodynamic system. However, this starting point is not trivial: the idea of *preparing* a system in a given macrostate is complex, and involves the notion of measurement. As we said in Chapters 3 and 5, classical mechanics is *a theory without an observer*, in the sense that terms such as measurement, observation and the like are not primitives of the theory and are accounted for in mechanical terms. Therefore, we need to describe the preparation of an initial macrostate by way of measurement in purely mechanical terms. We undertake this in the present chapter.

How does a system come to be in $[M_0]$? On the one hand, a system is always in a *microstate*, and the macrostate $[M_0]$ is a set of microstates, between which O does not distinguish. On the other hand, O knows with certainty that the system is in the macrostate $[M_0]$ rather than in any other macrostate. In the present chapter we shall explain the way in which this interplay between information (concerning the macrostate of the system) and ignorance (with respect to the microstate within that macrostate) comes about, as a consequence of the set up called *measurement*.

The idea that there may be a linkage between the notion of measurement and the laws of thermodynamics was first pointed out in Leó Szilárd's 1929 paper, "On the decrease of entropy of a thermodynamic system by the intervention of an intelligent being",[1] in which Szilárd analyzed the entropic changes in an ideal measurement interaction. Szilárd was correct in observing that the right way to understand the

---

[1] In Leff and Rex (2003, pp. 110–119).

notion of classical measurement is in statistical mechanics, where the notion of macrostates can be employed. However, his conclusions, regarding the particular way in which entropy changes during measurement, are wrong, and we will show how entropy changes during measurement. As the title of his paper shows, Szilárd was convinced that the idea of an *observer* (in his terms, an *intelligent being*) is essential for the understanding of the way in which entropy changes during measurement. Indeed, our study of the notion of measurement has consequences concerning the status of the observer in classical mechanics.[2]

9.2 What is measurement in classical mechanics?

In classical mechanics a measurement is an interaction between an observer O, a measuring device D, and a measured system G, which *brings about* correlations between the post-measurement states of O and D and the pre-measurement state of G. That is, the states of O and D are not correlated with the state of G before the measurement interaction and become correlated with the state of G only after the measurement. This lack of correlation before the measurement can be interpreted (in the appropriate circumstances) as *ignorance* with respect to the state of G (the notion of ignorance involved here will be discussed below); and this ignorance is *eliminated*, and replaced by *information*, at the measurement interaction. (For simplicity we consider here only ideal interactions, which are non-disturbing – that is, the state of G is unaltered by the measurement – and which induce perfect correlations between the state of D and the state of G.)

However, there are two kinds of ignorance that need to be sharply distinguished. Suppose that for a given observer there is a maximally fine partitioning of the state space into macrostates. If the observer happens to know the actual macrostate of a system which belongs to this partitioning (that is, if the observer is correlated with that actual macrostate), then the remaining ignorance about the actual microstate of the system cannot be further eliminated. If, on the other hand, the observer happens to know that a system is in a union of such fine-grained macrostates, then this ignorance can be eliminated by further measurements. Both kinds of ignorance play a role in thermodynamics. It is the second kind of

---

[2] In Appendix B.3 we make some comments about the thermodynamic *in*significance of measurement in quantum mechanics.

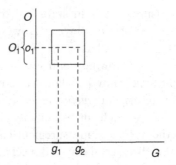

Figure 9.1 The condition of indistinguishability for measurement

ignorance, namely ignorance regarding maximally fine-grained macro-states, that we focus on in this chapter.

This elementary characterization of measurement, as bringing about correlations that did not exist prior to it, entails that in classical mechanics there are *no* microscopic measurements: a classical account of measurements has to be carried out in terms of macrostates.[3] The reason is this: suppose that we set up a measurement in which an observer O, possibly by means of an interaction with a measuring device D, will ascertain whether a particle G is located in the right-hand side or the left-hand side of a box. Since classical mechanics is deterministic, given the initial microstate of $O + D + G$ one can predict with certainty the final microstate of $O + D + G$, and so no ignorance is involved, and therefore the measurement is pointless, or even meaningless, under such a description. The whole point of measuring G's state by O (possibly using D) is to *gather* information about G's state, which O *lacks* prior to the measurement. The kind of information that O lacks prior to the measurement and O gathers by the measurement has to be of a kind that O cannot gather only by calculation, and the only way of gathering it is by carrying out the measurement. Therefore, given that the classical dynamics is deterministic, in order even to describe the pre-measurement ignorance of O about the state of G, the pre-measurement state of O must be consistent with at least two states of G, between which O cannot (at the pre-measurement state) distinguish. An example of this is shown in Figure 9.1, where O, which is in microstate o_1, cannot distinguish between the microstates

---

[3] By contrast, in quantum mechanics, measurement can be described by the microscopic dynamics provided that it is indeterministic.

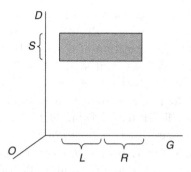

Figure 9.2 A pre-measurement macrostate

g_1 and g_2 of G. A set such as $\{g_1, g_2\}$, as we have already learned in Chapter 5, is a *macrostate* of G relative to O.[4]

The above elementary characterization of a measurement contains three conditions which are necessary and, together, sufficient in order for an interaction to be a measurement. We now spell out these three conditions in more detail.

(A) Pre-measurement

Initially, before the measurement, O is in some *microstate O_s*, the measuring device D is in some designated macrostate $[S]$ (relative to O), and G is in the macrostate $[L+R]$ relative to O.[5] This pre-measurement situation is illustrated in Figure 9.2, where the horizontal axis represents the states of G, the vertical axis represents the states of D, and, as usual, O is perpendicular to the page. The states of these three elements of this set up are correlated as follows. (i) The set $[S]$ is correlated with the microstate O_s of O, so that O experiences D as being in $[S]$. (ii) The dynamical set up is such that when D is in $[S]$, G is

---

[4] Price (1996) criticizes an idea he dubs *μInnocence*, according to which correlations between systems is something that comes about *following* their interaction: that is, the correlation come into existence only after the interaction and does not exist before. However, in classical micromechanics there is no μInnocence. Only at the macro level can correlations come about at a certain point of time, as is the case in our notion of classical measurement.

[5] Our account of *measurement* is consistent with Albert's (2000, Ch. 6) account of the notion of *record* (though the latter is not necessary for the former). Roughly, in Albert's account, a record consists of three consecutive events: first, at some time t_0, the measuring device has to be in some Ready state; second, at some later time t_1, the measuring device has to interact with the measured system such that the state of the measuring device will become correlated with, and reflect, the state of the measured system; and finally, the post-interaction state of the measuring device at t_1 has to remain unchanged until the later time t_2 at which the observer consults the measuring device and infers from it the state of the measured system at t_1.

somewhere in the box rather than anywhere outside the box; in other words, G is in the macrostate $[L+R]$.[6] (iii) The microstates of D in $[S]$ are *not* correlated with the position of G in the box, so that O cannot read off from the pre-measurement state $[S]$ of D whether G is in the left-hand side or the right-hand side of the box. Conditions (ii) and (iii) together mean that the pre-measurement macrostate of $D+G$ (relative to O) is $[S, L+R]$, as illustrated in Figure 9.2.

The status of the macrostate $[S, L+R]$ of $D+G$, and, in particular, the claim that the macrostate of G is $[L+R]$, needs some clarification. It is an uncontroversial fact that according to the ontology of classical mechanics, G is in a definite position in the box and, *ipso facto*, there is is a matter of fact concerning whether its actual position is in the left-hand side (L) or the right-hand side (R) of the box. Consequently, one might wonder why we claim that the macrostate of G is $[L+R]$, rather than either $[L]$ or $[R]$. To appreciate the gist of our analysis of this situation, notice that there is also a matter of fact concerning whether G is in either the upper or lower half of the box and, by the same logic, one might say that G's macrostate is either $[UP]$ or $[DOWN]$, instead of the entire region of $[L+R]$; and similarly, any partitioning of the $[L+R]$ would make the same sense, given that the actual state of G is represented by a point in that region; and so it *might* seem that the assignment of any of the regions $[L]$, $[R]$, $[UP]$, $[DOWN]$, etc. to G is *arbitrary*. But this is a mistake. In our approach, as explained in Chapter 5, the assignment of a macrostate to a system, such as the assignment of $[L+R]$ to G, is objective and physical, since it is a consequence of the physical interactions between systems and of the accessible region which is determined by external constraints and objective limitations. To say that G is in the macrostate $[L+R]$ (at the pre-measurement time) is to say that the microstate of G (at that time) is a member of a set of points, called $[L+R]$, all of which are correlated with the single microstate O_s of O, and this one-to-many correlation is an objective feature of O, of G, of the interaction between them, and of the rest of the world. Any claim to the effect that one can equally well assign to G any of the macrostates $[L]$, $[R]$, $[UP]$, $[DOWN]$, etc., is a mistake about the *objective* state of affairs in the world.

---

[6] This fact has been established by a prior measurement, which is part of the preparation of $D+G$ for our subsequent experiment. Also, we take it that when D is in $[S]$ it is ready to measure the $[L]$ or $[R]$-location of G within the box (for example, D is powered on, put in the right place, right orientation, etc.).

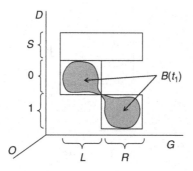

Figure 9.3 Split

(B) Split

The measurement interaction is such that the trajectories, which start out in the macrostate $[S, L + R]$ at the initial time t_0, evolve in such a way that at time t_1, by the end of the measurement evolution, the dynamical blob $B(t_1)$ partially overlaps with two macrostates: $[0, L]$ and $[1, R]$, as illustrated in Figure 9.3. This dynamical evolution brings about correlations between the macrostates $[0]$ and $[1]$ of D, corresponding to the two possible outcomes of the measurement, and the $[L]$ and $[R]$ macrostates of G: all the microstates in the blob $B(t_1)$ which are in the region $[0]$ of D are also in the region $[L]$ of G; and similarly for $[1]$ of D and $[R]$ of G. We call this kind of evolution *split*, to emphasize the overlap with the two macrostates; in general, however, the parts of the blob remain topologically *connected* (assuming that the topological structure of the trajectories at the pre-measurement state is connected[7]).

In Figure 9.3, the Lebesgue measure of the union of $[0, L]$ and $[1, R]$ is greater than the Lebesgue measure of $[S, L+R]$; thus because of Liouville's theorem (by which the Lebesgue measure of the dynamical blob is conserved), the dynamical blob $B(t_1)$ only *partly* overlaps with the two macrostates. In the special case where the relevant measure of the union of $[0, L]$ and $[1, R]$ happens to be equal to the measure of $[S, L+R]$, the blob will *completely* overlap with $[0, L]$ and $[1, R]$. This case is illustrated in Figure 9.4. The latter case often appears in the literature, possibly because of its simplicity;[8] but the case illustrated in Figure 9.3 is more general.

Notice that at the split stage, the relevant macrostates of G are $[L]$ and $[R]$ (and no longer the single macrostate $[L+R]$) because O is able to

---

[7] See discussion of this point in Hemmo and Shenker (2010).
[8] For example Bennett (1982); but Fahn (1996) rightly argues that this is not the general case. Our conceptual framework is different from Fahn's.

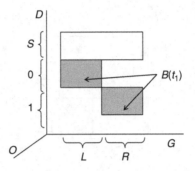

Figure 9.4 Special case of exact overlap between blob and macrostates

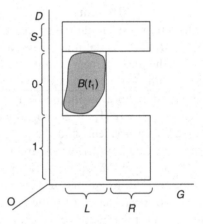

Figure 9.5 Split is necessary for measurement

distinguish between these sets of microstates of G, by looking at D. Once again, the correlations between O, D, and G are objective and physical since they are brought about by the interactions between these systems, and expressed by the accessible region which is affected by external constraints and other objective limitations.

The split stage is a *necessary* element of a classical measurement since it forms the bridge between the pre-measurement stage of ignorance and the later stage in which the measurement outcome comes about (as we explain shortly). To see why, consider Figure 9.5, in which the Lebesgue measure of *each* of the macrostates $[0, L]$ and $[1, R]$ is equal to (or even larger than) the Lebesgue measure of $[S, L+R]$, so that an evolution from $[S, L+R]$ to, say, $[0, L]$ (as in the figure) satisfies Liouville's theorem. An evolution from $[S, L+R]$ to $[0, L]$ only would not be a measurement since in that case one would be able to read off from the *initial* macrostate of $[S]$ of D

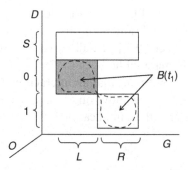

Figure 9.6 Measurement outcome

the fact that the particle G is in region $[L]$ of the box (and similarly for an evolution from $[S, L+R]$ to $[1, R]$). In other words, in such a case there would be no uncertainty at the initial time t_0 about the position of G, and the evolution described in the figure would not increase the amount of information that one can read off from D about the state of G. This is contrary to the very idea of what measurement is all about.

(C) Outcome

By the end of the measurement, the actual microstate of $D+G$ is either in the macrostate $[0, L]$ or in the macrostate $[1, R]$; and the macrostate in which the actual microstate of $D+G$ happens to be is the macroscopic outcome of the measurement. Let us now add O's state to this picture. The macrostates $[S]$, $[0]$, and $[1]$ of D are sets of microstates of D that are correlated with three microstates of O, let us call them o_S, o_0, and o_1, and so the actual microstate of $D+G$ determines whether O's microstate will be o_0 or o_1. Since, by our assumption (discussed in Chapter 5), O's microstate gives rise to O's experience, by the end of the evolution O has either the experience that D is in $[0]$ or the experience that D is in $[1]$; and from this, together with an acquaintance with the correlations between D and G (acquaintance which induces O to use D as a measuring device of the position of G), O infers whether G's actual position is in $[L]$ or $[R]$. Thus, the final state is this: either O is in o_0 and has the experience that $D+G$ is in $[0, L]$, or O is in o_1 and has the experience that $D+G$ is in $[1, R]$.

Notice (and this will turn out to be important) that since O's final microstate is correlated with the *entire* macrostate $[0, L]$ (or $[1, R]$, as the case may be), O assigns the entire macrostate $[0, L]$ to $D+G$, so that O describes the final state as in Figure 9.6. (In this figure, we indicate the dynamical blob by a dashed line since the blob no longer plays a role: O knows that $D + G$ are in $[0, L]$ and no probabilities are involved.) Whether or not the blob $B(t_1)$ covers $[0, L]$ partially (as in Figure 9.3) or completely (as in Figure 9.4) is immaterial here.

These are the three conditions that an interaction must satisfy in order to be a measurement. We now move on to study some aspects of them in more detail, and examine their implications.

9.3 Collapse in classical measurement

By the end of the measurement, O says that the measurement has an outcome, and that this outcome means that the macrostate of $D + G$ is *either* $[0, L]$ *or* $[1, R]$; at the same time the blob covers regions in the ODG space of which the projection is on *both* macrostates $[0, L]$ and $[1, R]$. These two consequences are compatible. In order to understand this compatibility, it is crucial to distinguish between the dynamical evolution of the blob as well as the evolution as it is described in terms of macrostates. We discuss this matter next.

The evolution of the blob is determined by the equations of motion and is subject to Liouville's theorem, according to which the Lebesgue measure of the blob must be conserved at all times. In particular, during the measurement interaction the evolution takes the bundle of trajectories from the initial volume $(o_s, S, L + R)$ in the state space of $O + D + G$, to the union of the regions $(o_0, 0, L)$ and $(o_1, 1, R)$ in this state space.

By contrast, the evolution of $D + G$ in terms of their macrostates is not directly subject to Liouville's theorem. The idea of evolution of macrostates needs some explanation.

The actual state of $D + G$, according to mechanics, is their microstate. The macrostates of $D + G$ are the only way in which O experiences their states, owing to the one-to-many correlation between O's microstates and the microstates of $D + G$. In this sense, it is the macroscopic evolution of $D + G$ that is experienced by O. The evolution of the microstate of O is given the *projection* of the blob on the O axis. And since the experience of O is determined by O's microstates, we can say that *the macroscopic evolution of $D + G$ is the microscopic evolution of O*, particularly in cases where O interacts with $D + G$. This has the consequence that the projected evolution of O's microstate is effectively stochastic. In this sense the experience of observers is *objectively stochastic* despite the underlying deterministic evolution of the universe.

Consider now the evolution of O's microstate as a result of the measurement interaction. At the pre-measurement stage, at time t_0, when O knows that $D + G$ is in $[S, L + R]$, O cannot predict whether $D + G$ will end in $[0, L]$ or $[1, R]$. Calculating the evolution of the blob from t_0 to t_1 is not going to help O to make such a prediction since at t_1 the blob spreads over the two possible final macrostates (as in Figure 9.3). Of course, O can calculate the probabilities of transition to the macrostates $[0, L]$ and $[1, R]$ by the Probability Rule, but cannot predict the actual outcome. For this O needs to carry out the measurement. At t_1, however, when the measurement is completed, O can assign one of these macrostates to $D + G$, and the reason is that the actual microstate of O becomes one of the two: either o_0 or o_1.

We call this assignment of macrostates to $D + G$ the *classical collapse* of the blob $B(t)$ onto the observed macrostate.[9] This classical collapse involves two parts: (I) detection of the actual macrostate, and (II) expansion over the detected macrostate. We now go on to explain these two parts.

(I) **Detection of the actual macrostate**. Suppose that the actual microstate of $D + G$ is in $[0, L]$ (and not in $[1, R]$). The correlations between O and $D + G$ are such that in this case the microstate of O is o_0. In this microstate, O has the experience that the macrostate of $D + G$ is $[0, L]$. Now, having this information about $D + G$, when O turns to make further predictions concerning the state of $D + G$, it would be pointless to carry out calculations on the basis of the full blob (that started out in $[S, L + R]$ at t_0 and evolved as in Figure 9.3). It makes much more sense to chop off, as it were, the part of the blob that is in $[1, R]$ and take into account, for all practical purposes, only the part that is in $[0, L]$. Notice that what this chopping off does is compatible with Liouville's theorem, since it does not mean that the trajectories end somehow at $[1, R]$, but only that their subsequent part is *ignored*.

(II) **Expansion over the detected macrostate**. In our example, O comes to know that the actual microstate of $D + G$ is in $[0, L]$. This is on the basis of direct observation which is, in general, not accompanied by calculating the evolution of the blob. Thus, in general O does not know which part of the macrostate $[0, L]$ overlaps with the blob $B(t_1)$ and which does not. In this sense, O experiences the outcome $[0, L]$ of

---

[9] Of course, we deliberately refer to the notion of collapse in quantum mechanics. But the classical notion is very different from the quantum mechanical one.

the measurement in terms of the *entire* macrostate [0, L]. This means that from the end of the measurement onwards, O carries out calculations *as if* the region representing the state of D + G were the entire macrostate [0, L], rather than the region of overlap with the blob B(t₁). That is, O carries out future calculations *as if* the part of the blob B(t₁) that overlaps with [0, L], spreads out and fills the entire region of [0, L].[10] This spreading out of the blob is *compatible* with Liouville's theorem for reasons similar to those stated in (I) above.

It is interesting to note that since the classical collapse generally involves expansion, it actually reduces the amount of information that one has about the *history* of the measured system. This point will become crucial in the case of retrodiction and in the way that statistical mechanics can account for the past (see Chapter 10).

When O's microstate has become correlated with either [0, L] or [1, R], (I) and (II) take place simultaneously.

A consequence of the expansion in (II) is the dependence of the transition probabilities calculated by the Probability Rule on observational history, as discussed in Section 6.6.

9.4 State preparation

Until now, whenever we have described an evolution in terms of dynamical blobs we started with an initial macrostate, which we usually called [M₀]. However, this already assumes that the system of interest has been prepared in [M₀]; and the question is: how is such a preparation carried out? In a nutshell, the answer is that the only way to control a system's state is to determine the external constraints on it, which are, in general, consistent with several macrostates: the macrostate itself cannot be determined directly. And so, in order to prepare a system in a given macrostate we need to determine the external constraints, and then to observe the system, *measure* its macrostate, and wait until it *spontaneously* reaches the desirable macrostate (and, therefore, the measurement is an *essential* part of state preparation). We emphasize that there is *no* other way to prepare a macrostate. In particular, investing work is just one way of determining external constraints and within them, generally, several macrostates are

---

[10] This idea, of expansion of the part of the blob that overlaps with the observed macrostate over that macrostate, is inspired by Gibbs's idea of coarse-graining.

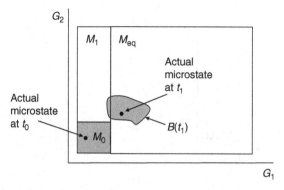

Figure 9.7 Preparing a system in equilibrium: dynamical evolution

possible. There is no way to bring about one of them rather than the others; the only way to prepare a macrostate is to wait until the system reaches the desired macrostate spontaneously, and discover this fact by measurement. Let us now describe this idea in some detail.

Consider, to begin with, a kind of preparation which is of special interest in the context of underwriting thermodynamic by mechanics, namely, preparing a thermodynamic system (such as a gas in a container) in its equilibrium macrostate. Consider Figure 9.7, in which G_1 and G_2 are two representative degrees of freedom of the gas G. The macrostate $[M_0]$ is the initial macrostate of G, and all of the trajectories that start in $[M_0]$ at t_0 evolve to the dynamical blob $B(t_1)$ at t_1. At t_0 the observer (possibly aided by a measuring device) is unable to say whether at t_1 the gas will actually be in $[M_{eq}]$ or in $[M_1]$; this is why probabilities are assigned to these two options, using the Probability Rule. Since the observation capabilities of O correspond to the partitioning of the state space into macrostates, the only way for O to know which is the actual macrostate of G at t_1 is to observe G at t_1. Suppose that O does so. The correlations between the microstates of O and G are such that if G's microstate happens to be in $[M_1]$, O's microstate will be o_1, and if G's microstate happens to be in $[M_{eq}]$, O's microstate will be o_{eq}, which is the case described in Figure 9.7. Once O's microstate is o_{eq}, O has the experience that G is in $[M_{eq}]$, and can treat G accordingly. Of course, the way to prepare the gas in $[M_1]$ given that it is initially in $[M_0]$ is exactly the same; but if the probability for evolving to $[M_1]$ is small, preparing the gas in $[M_1]$ in this method may take a long time. (An alternative method may involve investment of work.)

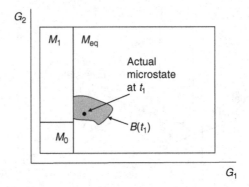

Figure 9.8 Preparing a system in equilibrium: detection

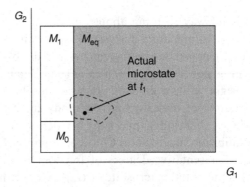

Figure 9.9 Preparing a system in equilibrium: expansion

Once O discovers that at t_1 G is in $[M_{eq}]$, two stages take place simultaneously. First, once O has detected the actual macrostate, O ignores the counterfactual case of G ending up in $[M_1]$; this case, which was a possibility (though with low probability) just a moment earlier, ceases to be a possibility. The part of the dynamical blob $B(t_1)$ which overlapped with $[M_1]$ is, as it were, chopped off, as in Figure 9.8. Notice, once again, that this chopping off does not violate Liouville's theorem since it is merely a step of *disregarding* the chopped off part for the purpose of future predictions.

The second stage (which happens simultaneously with the first) is a result of the fact that O is not aware of the details of the blob $B(t_1)$, but only of the fact that G is in $[M_{eq}]$. The expression of this fact in terms of the state space is that, from t_1 onwards, O treats G as being in the entire macrostate $[M_{eq}]$, as in Figure 9.9. The chopped off blob spreads out, as it were, all over the macrostate.

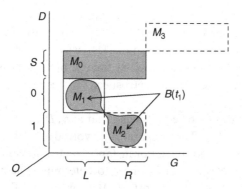

Figure 9.10 Preparation of an initial macrostate

At this stage O can carry out operations on G, using the information that the macrostate of G is $[M_{eq}]$. Incidentally, preparing a system in its equilibrium state may not seem to be very useful, since not much work (if any) can be produced at that state. However, this process is of key importance in our framework. The observer O can, for instance, weaken the constraints on G, for example by opening the container and letting the gas expand into a larger container. Once the container is opened, $[M_{eq}]$ is no longer the equilibrium state. And so, what we have described above is the way to prepare a system in an initial macrostate, such as in Figures 7.1, 7.2 and 7.3. In all these figures, a system has been prepared in an initial macrostate $[M_0]$, and we can now add that this $[M_0]$ contains (in the sense of (II) above) part (in the sense of (I) above) of the dynamical blob that originated from an earlier macrostate. We now know that O experienced that the system was in $[M_0]$ rather than in some alternative macrostate, and therefore (I) ignored the rest of the blob outside $[M_0]$ and (II) treated the system as being anywhere in $[M_0]$.

In order to see some more important aspects of the notion of preparation and its connection with measurement, consider again the case in which an observer O uses a measuring device D in order to measure the macrostate of G, as illustrated in Figure 9.10. The pre-measurement macrostate of $D + G$, at t_0, is $[M_0]$; $[M_1]$ and $[M_2]$ are the two possible post-measurement macrostates of $D + G$; and $[M_1]$ turns out to be the actual outcome of the measurement at t_1. In the figure, the shaded areas are those covered by the dynamical blob at the different stages of this evolution: $[M_0]$ at t_0, and $B(t_1)$ (partly overlapping with $[M_1]$ and $[M_2]$) at t_1. The regions surrounded by continuous lines in the figure are the actual macrostates at the different times: $[M_0]$ at t_0 and $[M_1]$ at t_1 (assuming that this is the actual outcome of the measurement, as in the above example). The areas surrounded by dashed lines are macrostates that did not contain the actual microstate during the time of interest: for example, $[M_2]$ was a possible outcome of the measurement as far as O could tell at time t_0, but this possibility did not materialize, and after t_1, in retrospect, it became a counterfactual macrostate.

Let us see what is involved in saying that at t_0, $D + G$ has been *prepared* in the macrostate $[M_0]$. This preparation involved an earlier measurement, one that took place prior to t_0. For instance, such a previous measurement could have been carried out in order to determine whether the particle G is in the box (which is macrostate $[M_0]$) or outside the box (which is macrostate $[M_3]$ in the figure). The preparation of G in $[M_0]$ was completed when at t_1 the blob collapsed onto the entire macrostate $[M_0]$. Operationally, this collapse means (for instance) that once G was found in $[M_0]$ the lid of the box was closed, and the experiment (of measuring whether G is in the left-hand side or the right-hand side of the box) could begin. The other possible outcome of the prior measurement – namely $[M_3]$ – was *ignored henceforth* by O. The very notion of preparing a system in a given macrostate contains the discovery that the actual microstate of the world is in one macrostate rather than another, and consequently, the *decision to follow* only the trajectories that start out in the actual macrostate. At time t_0, O prepared $D + G$ in $[M_0]$ and, after that time, O ignored the counterfactual $[M_3]$ altogether. At t_1 the same idea is repeated: O prepares $D + G$ in $[M_1]$, and henceforth ignores $[M_2]$ altogether.

9.5 The shadows approach

Often in the literature,[11] it seems that the counterfactual outcome, the one that did not take place, such as $[M_3]$ prior to t_0, or such as $[M_2]$ at t_1, in Figure 9.10, is not ruled out, and the evolution is described as in Figure 9.11. (The blob does not usually appear in the literature. We have marked it with dotted lines just to keep contact with the previous figures and appreciate the difference. The observer O does not figure in this approach either, and therefore we have omitted its axis from the figure.) We call this the *shadows approach*, since it retains the shadows

---

[11] For example Bennett (1982).

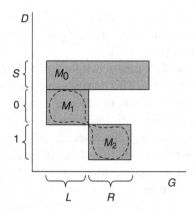

Figure 9.11 The shadows approach

of the past, that is, the counterfactual measurement outcomes that did not materialize.

This approach is somewhat *incoherent* in the way in which it treats the counterfactual outcomes, for while it discards $[M_3]$ (of Figure 9.10) by taking $[M_0]$ to be the prepared state of the system at t_0, it retains $[M_2]$ (as in Figure 9.11), and thus is unable to see that $[M_1]$ is a newly prepared macrostate. Consequently, the similarity between the status of $[M_3]$ and $[M_2]$ is missed; but if $[M_2]$ is retained, which is a counterfactual shadow of the past state of knowledge of O, $[M_3]$ ought (for the sake of coherence) to be retained as well. And if shadows ($[M_2]$ or $[M_3]$, as the case may be) are retained, the notion of state preparation becomes useless.[12]

9.6 Entropy

In the scenario of measurement in Figures 9.7, 9.8, and 9.9, entropy *increases* in the measurement, since the Lebesgue measure of the final macrostate $[M_{eq}]$ is larger than the Lebesgue measure of the initial macrostate $[M_0]$ (assuming that the Lebesgue measure is the v-measure of entropy). For exactly the same reason, the entropy of $D + G$ *decreases* during the measurement described in Figures 9.2, 9.3, and 9.6. The first

---

[12] One way of explaining the origin of this mistake is to see it as a lack of distinction between dynamical blobs and macrostates. A motivation for the shadows approach might come from quantum mechanics, in which case, according to the projection postulate, a measurement can be thought of as a preparation of a new quantum state.

lesson here is that classical measurement is compatible with any change of entropy, and that this change depends only on the v-measure of entropy of the initial and final macrostates.

It may be the case that as a matter of fact we are surrounded by many examples of entropy decrease. This depends only on the relative measures of the macrostates involved in the measurements that we carry out whenever our senses interact with our environment. But this should not be surprising: after all, entropy expresses an observer's ability to exploit energy, and we can better exploit the energy in systems around us whenever we have more accurate information about their microstates. And it is precisely this that we gain by measurement.

Szilárd[13] famously claimed that measurement necessarily involves a decrease of entropy by $k\log 2$ in $D + G$ (he had in mind something like Figure 9.4), and must be accompanied by an increase of entropy elsewhere in the universe. It is currently a prevalent opinion that Szilárd's claim is wrong, since measurement does not involve an entropy decrease in $D + G$; but this opinion rests on ideas concerning the thermodynamics of computation, which we address in Chapter 12, and which we think are fundamentally mistaken. Our reasons for rejecting Szilárd's claim are very different. We think that: (a) Szilárd was right in noticing that measurement *can* bring about a decrease of entropy; but (b) he was wrong in thinking that such a decrease is *necessary*; and (c) he was wrong in concluding that such a decrease, whenever it occurs, must be accompanied by an increase of entropy elsewhere in the universe. Although our reasons for thinking that may be clear by now, let us spell them out briefly.

The reason for (a) and (b) is that as we saw above, during measurement entropy can change in any way: it can increase (Figures 9.7, 9.8, and 9.9), or decrease (Figure 9.3 and 9.4), or remain unchanged.[14] The only limitation on the change of entropy during measurement is an indirect consequence of Liouville's theorem: the Lebesgue measure of *all* the macrostates of $D + G$ associated with the measurement's outcome taken together (and in general there can be more than two outcomes), multiplied by the Lebesgue measure of the suitable region along the O degree of freedom, cannot be smaller than the Lebesgue measure of the initial

---

[13] Szilard (1929).

[14] Notice that this has nothing to do with the fact that measurement is a logically reversible operation. We discuss the entropy of the physical implementation of logical operations in Chapter 12.

macrostate of $D + G$ – multiplied, again, by the Lebesgue measure of the suitable region along the O degree of freedom.

The reason for (c) is twofold. First, the entropy decrease in measurements does not stand in contradiction to the statistical mechanical counterparts of the Second Law of Thermodynamics and the Law of Approach to Equilibrium, as stated in Chapter 7. This decrease is only due to the details of the correlation between O and $D + G$ which brings about a partitioning of the state space into macrostates. Second, Szilárd's claim rests on his belief that the statistical mechanical counterparts of the Second Law of Thermodynamics and the Law of Approach to Equilibrium are universally true,[15] but as we saw in Chapter 7, there is as yet no general proof of these laws from mechanics. Indeed, we think that such a proof cannot be forthcoming, since there is a counter example to it, in the form of Maxwell's Demon, discussed in Chapter 13. Therefore, *contra* Szilárd, a consequence of the above analysis is that measurement can decrease the total entropy of the universe.

Szilárd's discovery was nevertheless remarkable and path-breaking because he pointed out the role of the notion of an observer (in his term: *intelligence*) in classical statistical mechanics. Indeed, the change of entropy during measurement is a consequences of the way in which $D + G$ interact with the observer O, an interaction which brings about the phenomenon of macrostates. In an important sense, our notion of experience parallels Szilárd's notion of intelligence.

9.7 Status of the observer

In Chapter 5 we saw that the very notion of a macrostate assumes an observer: the actual state of the universe is a microstate, and sets of microstates of the observed system G and the measuring device D form macrostates because of their objective and physical correlation with the microstates of the observer O. This status of the observer is crucial for the idea of a classical measurement: the initial macrostate evolves as in Figure 9.3, in which there is no indication for the measurement's outcome; and it is only by virtue of the fact that an observer has an *experience* that we can explain the fact that measurements have outcomes.

---

[15] See Earman and Norton (1999).

The status of the observer in our approach in classical statistical mechanics has a striking counterpart in a famous argument by John von Neumann in his *Mathematical Foundations of Quantum Mechanics*.[16] Although the context of von Neumann's discussion is the measurement problem in quantum mechanics, we believe that his insight is relevant in the classical context as well. Here is what von Neumann says about the inevitable notion of experience:

> We wish to measure a temperature. If we want, we can pursue this process numerically until we have the temperature of the environment of the mercury container of the thermometer, and then say: this temperature is measured by the thermometer. But we can carry the calculation further, and from the properties of the mercury, which can be explained in kinetic and molecular terms, we can calculate its heating, expansion, and the resultant length of the mercury column, and then say: this length is seen by the observer. Going still further, and taking the light source into consideration, we could find out the reflection of the light quanta on the opaque mercury column, and the path of the remaining light quanta into the eye of the observer, their refraction in the eye lens, and the formulation of an image on the retina, and then we could say: this image is registered by the retina of the observer. And were our physiological knowledge more precise than it is today, we could go still further, tracing the chemical reactions which produce the impression of this image on the retina, in the optic nerve tract and in the brain, and then in the end say: these chemical changes of his brain cells are perceived by the observer. But in any case, no matter how far we calculate – to the mercury vessel, to the scale of the thermometer, to the retina, or into the brain, at some time we must say: and this is perceived by the observer. That is, we must always divide the world into two parts, the one being the observed system, the other the observer. In the former, we follow up all physical processes (in principle at least) arbitrarily precisely. In the latter, this is meaningless. The boundary between the two is arbitrary to a very large extent . . . That this boundary can be pushed arbitrarily deeply into the interior of the body of the actual observer is the content of the principle of the psycho-physical parallelism – but this does not change the fact that in each method of description the boundary must be put somewhere, if the method is not to proceed vacuously, i.e., if a comparison with experiment is to be possible. Indeed, experience only makes statements of this type: an observer has made a certain (subjective) observation; and never any like this: a physical quantity has a certain value.[17]

We suggest reading this specific quotation from von Neumann as if it had been written directly about *classical* measurement and the role of the observer in *classical* statistical mechanics. The insight we take from von

---

[16] von Neumann (1932).
[17] von Neumann (1932), Ch. 6, pp. 418–420

Neumann is that measurement outcomes cannot be accounted for unless one appeals to the experience of an observer. As to the question of whether one can account for the experience of observers in a way that is compatible with physicalism (as discussed in Section 5.8), our answer is that it may be done.[18]

---

[18] Although there is still a long way to go.

10

The past

10.1 Introduction

Normally, a theory is tested against experience by comparing its predictions with observations: after all, one of the main aims of science is to come up with good predictions. In many cases, we are much less interested in *retro*dictions. One reason, perhaps, is that we already know what happened in the past, at least to some extent, through our memories and through various records. Another reason may be that we cannot influence the past. The only thing that we can do about the past is to remember it, or reconstruct it from records, or retrodict it, and the like. Of course we can retrodict the past on the basis of our theories or we can conjecture which events occurred in the past and then derive from these conjectures predictions for later times (which are still in the past). But, in both cases, our retrodictions and predictions can be tested only against our memories and records of the past.

At the same time, it is natural to expect a theory which provides good *pre*dictions to provide also good *retro*dictions: in a sense, the ability to provide retrodictions that match our memories and records is a reasonable way to test a theory. In addition, there are facts about the past about which we have no memories and records, and it would be good if our theories entailed these facts. Important examples here are theories about the beginning of the universe and about the origin of life. But in order to trust our theories concerning these past events we need to be sure that our theories are good at retrodicting, not only at predicting. The problem is that *if* (or to the extent that) statistical mechanics is *good* at predicting the thermodynamic regularities, *then* (to the same extent) it provides *false* retrodictions – that is, it provides retrodictions of past events that are in contradiction with our memories and records. In this chapter we explain this problem and discuss its solution.

10.2 The problem of retrodiction

Consider a system G (represented by two of its degrees of freedom, G_1 and G_2) that is in the macrostate $[M_1]$ at the present time t_0, as illustrated in Figure 10.1. The dynamical evolution is such that the blob in the future time t_1 is as described in the figure, and using the Probability Rule we predict that G will evolve so that in the future time t_1 it is highly likely to reach the higher-entropy macrostate $[M_2]$.

We now further assume that at the present time t_0 we *remember* that G was prepared in the past time t_{-1} in some low-entropy macrostate $[M_0]$. Two questions that are closely linked (as will turn out later) arise. First, *the reliability question*: is this memory of ours reliable? Can we infer from the fact that we experience this memory in the present t_0 that, indeed, at t_{-1} G was in $[M_0]$? And second, *the retrodiction question*: can we infer, or retrodict, the contents of this memory from the information that the present macrostate of G at t_0 is $[M_1]$, using the principles of statistical mechanics? If the answer to the retrodiction question is positive, it may support a positive answer to the reliability question.

However, the answer to the retrodiction question in statistical mechanics is negative, as we will explain in what follows. And the way to overcome this problem is by postulating a positive answer to the reliability question; we shall call this positive answer the *Reliability Hypothesis*,

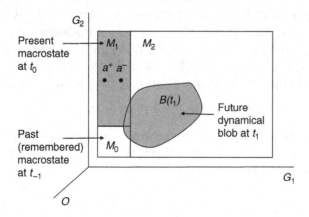

Figure 10.1 The minimum entropy problem

and argue that it is a good answer, and it is neither circular nor *ad hoc*. Let us begin with the problem of retrodiction.

At t_0, G is in some microstate in $[M_1]$, call it a^+. As explained in Chapter 4, retrodicting the past microstate of G is carried out by calculating the integral along the trajectory that starts out at the microstate a^+ from the present time t_0 toward the past time t_{-1}. This calculation is mathematically equivalent to calculating the integral from the present time t_0 to the future time t_1 (we assume here that $t_1 - t_0 = t_0 - t_{-1}$), together with replacing the present state a^+ by the state a^- in which the velocity is the reversal of the velocity of a^+ but the position is the same as in a^+.[1]

At this point we add the assumption that any two microstates that are the velocity reversals of each other belong to the same macrostate. As we said in Section 7.8, this assumption is reasonable in the context of thermodynamic macrostates: it is hard to imagine that reversing the velocities of, say, the molecules of a gas in a container will have an immediate effect on its macrostate (though of course its evolution will be affected by such a reversal). Although this assumption is neither necessary nor trivial, it is highly reasonable and is tacitly made in the literature, and we are going to adopt it in the present argument. By this assumption a^- is in $[M_1]$, and so anything we will say about a^+ by virtue of it being in $[M_1]$ will equally hold for a^-. In particular, since $[M_1]$ evolves to $B(t_1)$, both a^+ and a^- are in $B(t_1)$.

The partial overlaps between $B(t_1)$ and the macrostates are such that in our example illustrated in Figure 10.1, G is highly likely to evolve from $[M_1]$ at t_0 to the higher-entropy macrostate $[M_2]$ at t_1, and is less likely to evolve to the lower-entropy macrostate $[M_0]$ or to remain in $[M_1]$. And since both a^+ and a^- are both in $[M_1]$, both evolve to $B(t_1)$, and therefore a^- is as likely as a^+ to evolve to $[M_2]$. But a *pre*diction of the evolution of a^- from t_0 to t_1 is equivalent to a *retro*diction of the history of a^+ from t_0 back to t_{-1}, and so it turns out that if G is highly likely to evolve to the higher-entropy macrostate $[M_2]$, then it is also highly likely to have arrived from the higher-entropy macrostate $[M_2]$. A bit more formally:

$$P(M_2, t_{-1}|M_1, t_0) = P(M_2, t_1|M_1, t_0).$$

Thus, entropy is equally likely to increase by a transformation from $[M_1]$ to $[M_2]$ in *both* directions of time. That is to say, for example, we would be just as justified to infer that the cup of coffee in front of us was at room

---

[1] The forces have to be adjusted as well, but we do not go into this point here; we focus on the kinematic case.

temperature a few minutes earlier and has spontaneously warmed up as to infer that it will cool down in a few minutes.

Similarly, entropy is equally likely to *decrease* by a transformation from $[M_1]$ to $[M_0]$ in both directions of time:

$$P(M_0, t_{-1} | M_1, t_0) = P(M_0, t_1 | M_1, t_0).$$

The probabilities in this equation are affected by Liouville's theorem in that if (as in Figure 10.1) the entropy of the past macrostate $[M_0]$ is lower than the entropy of the present macrostate $[M_1]$, then a dynamical blob that starts out in $[M_1]$ cannot be fully contained in $[M_0]$, and so in this case our memory of $[M_0]$ can never be retrodicted with certainty. Obviously, this is the case we are interested in, since it implies that the Law of Approach to Equilibrium was true in the past. Thus we arrive at two conclusions about retrodicting the past, given that the Law of Approach to Equilibrium is true about our future. (I) Although we often feel that our memories of the past are certain, it is impossible to retrodict the past with certainty on the basis of mechanics alone. (II) Retrodiction gives high probability to a history that is different from our memories.

Notice that in both prediction and retrodiction, the direction of time that we have in mind is the one that forms an integral and indispensable part of the basic time-slices, and which defines the direction of velocity of particles, and forms part of the micromechanical equations of motion, as explained in Chapter 4.

This result is sometimes referred to as the *minimum entropy problem* since it shows that the present entropy of G is the minimal one, for *any* time taken as present. Since entropy is not constant, this outcome has an air of inconsistency; and it is in clear contradiction with our memories and records which are summarized by the time-asymmetric laws of thermodynamics.

The minimum entropy problem has the following consequence concerning the mechanical account of the phenomenon of memory: the mechanism of our memories *cannot* be a result of a calculation of macroscopic retrodiction that starts from the present macrostate and is based on calculating the backwards evolution of the dynamical blob. This is so both because the contents of our actual memories contradict the contents we would have if our memories were based on such macroscopic retrodictions; and because of the inconsistency of the idea that entropy is at its minimum at every point of time. This consequence is in line with some further conclusions regarding the mechanism of memory that suggest that memories supervene on microstates (rather than on macrostates; see below).

10.3 The Past Hypothesis: memory and measurement

A famous family of attempts to solve the minimum problem goes by the name of the *Past Hypothesis*. Here is Richard Feynman's version of the Past Hypothesis:[2]

> ... I think it necessary to add to the physical laws the hypothesis that in the past the universe was more ordered, in the technical sense, than it is today.[3]

By the term "more ordered" Feynman refers to a macrostate of lower entropy, given Boltzmann's macrostates presented in Chapter 5. The hypothesis is that as a matter of objective fact, the past microstates of the universe belonged to sets of thermodynamic macrostates that have lower entropy than does the present macrostate of the universe.

The Past Hypothesis is a complex idea, consisting of several independent ideas that should be examined. Two of the ideas that form part of the Past Hypothesis are the following.[4] (1) Our memories are reliable: they reflect the actual state of affairs that obtained in the past; we call this the *reliability hypothesis*. (2) The contents of our memories are that the entropy in the past was lower than it is at present: this is the *low entropy assumption*. Let us begin by showing how the reliability hypothesis works in our conceptual framework and why we think it is justified, and then we will address the low entropy assumption, and some other parts of the Past Hypothesis.

The Reliability Hypothesis actually means that macroscopic retrodictions are similar to measurements as described in Chapter 9. We trust our memories in a way that is analogous to the way that we trust our observations, and the theoretical understanding of these two cases is similar. To see this, compare the following two scenarios: one of measurement and the other of memory.

Start with measurement. Consider again the case in Figure 10.1, and suppose that the observer finds the system G at t_1 in macrostate $[M_2]$. In that case, as we explained in Chapter 9, a collapse takes place: the microstate of the observer gets correlated with the actual microstate that happens to be in $[M_2]$, and thereby the detection of the measurement

---

[2] See Albert (2000, Ch. 4) and Sklar (1973) for a variation on this idea.
[3] Feynman (1965, p. 116).
[4] Some versions of the Past Hypothesis include the Statistical Postulate and other assumptions from cosmology. For a critical discussion of the inputs from cosmology, see Earman (2006).

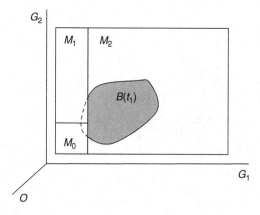

Figure 10.2 Detection of an outcome in measurement

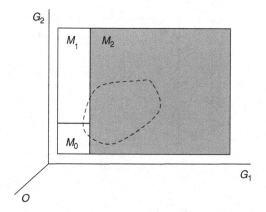

Figure 10.3 Expansion in measurement

outcome takes place. Therefore the observer ignores henceforth the parts of $B(t_1)$ that overlap with the other macrostates (see Figure 10.2). Consequently the observer treats the system as being in the entire macrostate $[M_2]$ (see Figure 10.3), which means that expansion takes place.

The case with retrodiction based on memory is exactly the same. Consider Figure 10.4, which gives an illustration of retrodiction. Notice that this figure is identical to Figure 10.1, except that the dynamical blob is called $B(t_{-1})$ instead of $B(t_1)$, since indeed, as we explained above in presenting the minimum problem, following the trajectories that start out in $[M_1]$ at t_0 towards their past leads to $B(t_{-1})$ which is identical to $B(t_1)$. Now suppose that at t_0 the observer *remembers* that at t_{-1} the system G was in macrostate $[M_0]$, and takes this memory to be reliable. In other

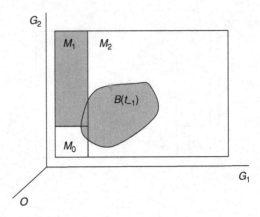

Figure 10.4 Retrodiction

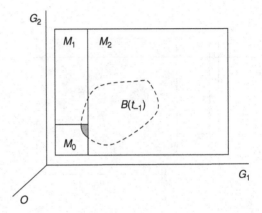

Figure 10.5 Detection in memory

words, the observer assumes that her memory is physically correlated with the fact that at t_{-1} the system G was in macrostate $[M_0]$. This memory entails a treatment that parallels the treatment of measurement. The microstate of the observer, in which its memory is stored, is correlated with the macrostate $[M_0]$ of G. Taking this correlation as reliable is a sort of retrospective detection. Consequently, the observer *ignores* henceforth the parts of $B(t_{-1})$ that overlap with the other macrostates (see Figure 10.5), and treats the system as being in the *entire* macrostate $[M_0]$. This means that an expansion takes place (see Figure 10.6)

Notice that the realization that memory is a kind of measurement of the past does not eliminate the minimum problem – unless we *assume* the Reliability Hypothesis. We now turn to address the justification for making this hypothesis.

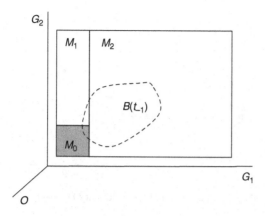

Figure 10.6 Expansion in memory

10.4 The Reliability Hypothesis

The problem with justifying the Reliability Hypothesis (above and beyond general sceptical qualms) is expressed in these famous words by Bertrand Russell:

> In investigating memory-beliefs, there are certain points which must be borne in mind. In the first place, everything constituting a memory-belief is happening now, not in that past time to which the belief is said to refer. It is not logically necessary to the existence of a memory-belief that the event remembered should have occurred, or even that the past should have existed at all. There is no logical impossibility in the hypothesis that the world sprang into being five minutes ago, exactly as it then was, with a population that "remembered" a wholly unreal past. There is no logically necessary connection between events at different times; therefore nothing that is happening now or will happen in the future can disprove the hypothesis that the world began five minutes ago. Hence the occurrences which are called knowledge of the past are logically independent of the past; they are wholly analysable into present contents, which might, theoretically, be just what they are even if no past had existed. I am not suggesting that the non-existence of the past should be entertained as a serious hypothesis. Like all sceptical hypotheses, it is logically tenable, but uninteresting. All that I am doing is to use its logical tenability as a help in the analysis of what occurs when we remember.[5]

Although, in our discussion, we are looking for physical rather than logical connections between present and past states, Russell's point is

---

[5] Russell (1921).

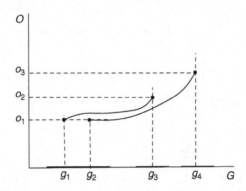

Figure 10.7 Russell's paradox and the Reliability Hypothesis

crucial precisely because the reliability of our memories cannot be derived from mechanics. In our conceptual framework, Russell's "five minutes ago" paradox comes about as follows. Suppose that at time t_1 an observer O remembers an event that took place at an earlier time t_0. For example, suppose that at t_0 the observer O observes a glass of cold water with a little ice in it, and at the same time O remembers that at an earlier time t_{-1} the glass contained only ice (and no water). This memory is brought about by the microstate of O's brain at t_0, call it o_1; see Figure 10.7. This memory microstate o_1 is compatible with two possible microstates of the rest of the universe at t_0: g_1 and g_2; we would say that g_1 and g_2 are in the same macrostate, indicated by the thick line segment on the G axis in Figure 10.7. Now the microstate o_1g_1 of $O + G$ is such that if we were to retrodict from it the microstates of O and G at t_{-1}, we would have found them in the microstate o_2g_3, in which O experiences a glass full of ice, and G is in a macrostate (indicated by a thick line segment on the G axis) that is compatible with such an experience. By contrast, the microstate o_1g_2 at t_0 is such that if we were to retrodict from it the microstate of $O + G$ at t_{-1}, we would have found them in the microstate o_3g_4, in which O experiences a glass full of water at room temperature. The latter case is Russell's "five minutes ago" paradox. Notice that false retrodictions of the kind just sketched are a consequence of the many-to-one correlations between O's memory states and the rest of the universe. In other words, it is a consequence of the notion of *macrostates*. Had the correlations between our memory states and the states of the rest of the universe been one-to-one, then Russell's paradox could not be formulated.

The Reliability Hypothesis is a denial of Russell's "five minutes ago" paradox, *by postulation*: it says that the actual cases we experience are like the evolution from o_2g_3 to o_1g_1 and not from o_3g_4 to o_1g_2, in Figure 10.7. The hypothesis does not involve a *proof* that that the "five minutes ago" scenario is false or even improbable; such a proof is impossible, since such trajectories are compatible with mechanics. Nevertheless, this postulation is not arbitrary, and there are good reasons to adopt it. We think that the strongest argument for adopting the Reliability Hypothesis is David Albert's line of reasoning which says, in a nutshell, that the very idea that there can be empirical data, an idea on which the whole of empirical science is based, relies on the reliability of our memories. Relinquishing the Reliability Hypothesis leads to a sceptical catastrophe and to relinquishing science as an empirically significant enterprise.[6]

Assuming the Reliability Hypothesis, we call trajectories such as the one from o_2g_3 to o_1g_1, along which our memories reflect the actual state of affairs in the past, *historical trajectories*; and the microstates that give rise to them, such as o_1g_1, are *historical microstates*. Trajectories such as the one from o_3g_4 to o_1g_2, and microstates such as o_1g_2 are called, respectively, *non-historical* trajectories and microstates.

Notice that reliance on memory, as illustrated in Figures 10.4, 10.5, and 10.6, does not involve any probability claims. Given that we remember the actual past state of the universe, there is no room for probabilistic claims about the past. When we remember a past event, we are certain that it has occurred. Of course, in many situations we do not have clear memories of the past but, in these cases, we say that we do not remember the past rather than assign probabilities to various possibilities. In this sense memory involves a yes–no detection, just like the detection of an outcome of a measurement. The detection itself is not probabilistic; probability plays a role only until the detection takes place. And moreover, once the detection has taken place it no longer matters what the probabilities were for its occurrence. For example, once one wins the lottery, it no longer matters that this event had low probability. And the same holds with respect to the detection of the highly probable event of losing in the lottery.

In addition to retrodictions based on actual memories, we can construct arguments based on a generalization of our memories. If, for example, we remember that all the systems we have observed had lower entropy in the past, we may inductively generalize this fact, and say that entropy is always lower as we go to the past. However, if we do not have

---

[6] For more details see Albert (2000, Ch. 4).

any memories about the past, and we do not generalize from our memories to the past state of affairs in other cases, then the method illustrated in Figures 10.4, 10.5, and 10.6 above cannot be applied: we cannot carry out the *collapse* step, since there is no detection. In that case, the best we can do is to rely on our theory, and use the Probability Rule with respect to the past; and in such a case indeed owing to the time-reversal invariance (see Figure 10.1) our retrodictions ought to be similar in nature to our predictions. We reiterate for emphasis: if we have *no* memories of the past, and we *do not* want to speculate that the unknown past resembled the known past (in particular, if we do not want to speculate that the thermodynamic empirical generalizations hold in unknown events and in the unknown past), then the best retrodictions we can make are precisely those in which the entropy change towards the past is the same as the entropy change towards the future: it increases and decreases with the same probabilities in both directions of time.

10.5 Past low entropy hypothesis

The Reliability Hypothesis is quite minimal since, by itself, it does not require that the contents of our memories be of macrostates of *low* entropy in the past. Although as a matter of *fact* we remember that thermodynamic systems evolved from low entropy to higher entropy, this is not a consequence of the Reliability Hypothesis as such (nor of the principles of mechanics, see Chapters 7 and 13). It is, for instance, compatible with mechanics and with the Reliability Hypothesis that instead of the memory, according to which the past macrostate is $[M_0]$ as illustrated in Figures 10.4, 10.5, and 10.6 above, one would have a memory according to which the past macrostate was $[M_2]$, and then a retrodiction of the state in Figure 10.4 would yield the macrostate in Figures 10.2 and 10.3. That is, in this case the historical high-entropy macrostate will be assigned high probability but this is a mere coincidence.

However, as a matter of fact our memories reflect the thermodynamic regularities, and we do generalize these memories to all relevant cases, and therefore believe that the Law of Approach to Equilibrium and the Second Law of Thermodynamics hold in all the past cases. And so the case described in Figures 10.4, 10.5, and 10.6 is generic. Let us examine some consequences of this fact.

According to the low entropy assumption, the macrostate $[M_1]$ at the present time t_0 is larger in v-measure (that is, the measure of entropy) than

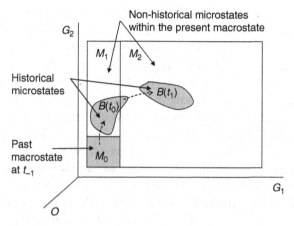

Figure 10.8 Historical and non-historical microstates

the macrostate $[M_0]$ at the past time t_{-1}. If the v-measure turns out to be the Lebesgue measure, then by Liouville's theorem the dynamical blob that started out in the past macrostate $[M_0]$ can overlap with only part of $[M_1]$, as in Figure 10.8. Now, when we observe the world around us and find it in the present macrostate $[M_1]$, by the low entropy assumption only the points in $[M_1]$ that are also in $B(t)$ are historical (in the sense defined above), and the rest of the points in $[M_1]$ are non-historical. In general, given the low entropy assumption, *most* (as measured by μ) of the points in the present macrostate $[M_1]$ do not originate from $[M_0]$ and therefore are non-historical. And as the evolution proceeds to $[M_2]$ at t_1, an even smaller part of the new present macrostate is historical.

And so on: as the universe evolves towards the future, more memories and records of the growing past are formed. This means that the subset of the historical trajectories gets *relatively* smaller and smaller as time goes forward. Moreover, if we assume that the constraints on the universe (or of the relevant subsystems of it) are released with time (for example, the size of the universe increases), then the size of the accessible region increases, and so the relative measure of the historical set of microstates becomes even smaller.

We emphasize that the assumption concerning low entropy expresses a *contingent* fact about the actual microscopic evolution of the universe, as it is seen in terms of thermodynamic macrostates. This assumption is not a theorem of mechanics: mechanics does not entail that the direction of time we call "past," the direction in which we have memories, is such that entropy is low in that direction.[7] Mechanics is also compatible with a case in which the actual microscopic evolution is such that entropy decreases.[8] The arrow of time is already built into the microscopic dynamics (as explained in Chapter 4) and therefore is not produced by the Past Hypothesis, and in particular it is not produced by the low entropy assumption. Given the primitive mechanical direction *of* time, the Past Hypothesis postulates a certain way in which entropy changes *in* time, and this way of change is not necessary, as far as classical mechanics is concerned.[9] As we saw, Tami's universe is perfectly consistent with the Newtonian equations of motion. In other words, the direction *in* time in which entropy increases may not be parallel to the direction *of* time built into the microscopic dynamics. The Past Hypothesis assumes or detects a direction of time and is given relative to it; it does not define nor bring about a direction of time.

10.6 Remembering the future

The Past Hypothesis assumes a direction of time and that it is meaningful to say that we remember the past, rather than the future. This assumption is consistent with the fundamental ontology of classical mechanics: as we saw in Chapter 4, each time-slice contains an inherent direction of time, without which velocities are not well defined. Given this direction of time, mechanics describes the evolution of systems in time. Given a direction of time in our actual universe, the notion of past is well defined and objective, and it is this past that features in our memories.[10] And assuming that the mental states of observers (such as memory) supervene on the physical

---

[7] This claim goes against a view (see for example Hawking 1988, Ch. 9) according to which the direction of memory is dictated by the direction of entropy increase. Since we think that the Law of Approach to Equilibrium and the Second Law are not necessarily universal, the realization that they do not determine the contents of our memories is very important. Hawking's argument is based on claims regarding entropic effects of measurement; but measurement is compatible with either entropy increase or decrease, as we saw in Chapter 9.

[8] And, of course, mechanics is also compatible with a different (non-thermodynamic) partitioning of the state space into macrostates, relative to which the actual microscopic trajectory may be either entropy-increasing or entropy-decreasing.

[9] There are good reasons to suppose that this is also the case with respect to quantum mechanics; but we do not pursue this here.

[10] This is why we are justified in postulating a Past Hypothesis rather than a future hypothesis; compare Price (1996): The past differs from the future in that we remember the past and do not remember the future. This is an objective physical fact although it is *not* entailed by the principles of mechanics (see below).

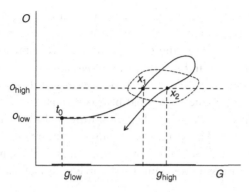

Figure 10.9 Remembering the future

microstates of these observers, it is reasonable to assume that the direction of time built into the physical microstates brings about (in one way or another) the experience of the direction of time.

However, the following example raises questions regarding the temporal order of our memories. Why is it that we remember only the past direction of time? Why, that is, do we not remember events that occur in the future? By this question we do not mean to ask how memory comes about; we will just show that no matter how our memory of the past comes about we could also remember events in our future as far as mechanics is concerned. Figure 10.9 describes an observer O that observes a system G, for example a gas in a box. The correlations between O and G are such that if G is in any microstate in the set g_{low}, in which the gas has low entropy, O is in the microstate o_{low}; and if G is in any microstate in the set g_{high}, in which the gas has high entropy, O is in the microstate o_{high}. At time t_0 the system G starts out in some microstate in the macrostate g_{low}, and O is in the microstate o_{low}. The dynamics brings about an evolution of $O + G$ so that at time t_1 the microstate of G is in the macrostate g_{high} and O is in the microstate o_{high}. At t_1 O's brain state o_{high} has two aspects: (i) O experiences g_{high}; (ii) O remembers that at t_0, in the past, the macrostate of G was g_{low}. This double role is, of course, the usual case. Call the microstate at t_1, in which O experiences g_{high} and remembers g_{low}, x_1.

As the system continues to evolve, Poincaré's theorem entails that at some later time t_2 the trajectory reaches a microstate x_2 which is in a close neighborhood of x_1. In particular, it could be that the projection of x_2 on the degree of freedom of O's brain which remembers the past is very close (or even identical) to the memory microstate at t_1. Now, if memory supervenes on (sets of) microstates, then the contents of O's memory in

x_2 must be the same as in x_1. In this case, there are only two possibilities: (i) either at t_2 O remembers everything up to time t_1, and forgets the events that took place between t_1 and t_2; or (ii) at t_1 O "remembers" events that took place between t_1 and t_2 – that is, O *"remembers" the future.*

Now, it is a fact that we remember only events in the past direction of time. But since mechanics is compatible (as we just showed) with trajectories in which we could remember the future, it follows that the claim that we remember the past rather than the future, or even that we remember only one direction of time rather than both, is not analytic in mechanics. On this issue we cannot but agree with Russell when he says:

> We all regard the past as determined simply by the fact that it has happened; but for the *accident* that memory works backward and not forward, we should regard the future as equally determined by the fact that it will happen.[11]

10.7 Problem of initial improbable state

As we said above, the low entropy assumption entails that the set of historical points has small relative measure (see Figure 10.8). This idea is sometimes presented from a different angle, in a way that makes it sound unnecessarily problematic. Rendered in terms of macrostates and dynamical blobs, this claim is this. Suppose we are now in macrostate $[M_1]$, and we do not make the low entropy assumption. In that case, in order to retrodict the past of the universe we need to apply the Probability Rule on the past blob, as in Figure 10.4. Now the measure of the overlap between the blob $B(t_{-1})$ in that figure and the macrostate $[M_0]$ is very small. Nevertheless, as a matter of fact, this was the actual macrostate. And the result is that *the actual past is improbable*. (The description of this idea in the literature is given in different terms, of course.[12])

We think that this result, which is an artifact of this way of presenting the above idea, is not problematic. For all we care, we need only *one microscopic trajectory* to agree with all our memories in order to have a full explanation of the history of the universe. We see no explanatory advantage in the requirement that the historical subset be of positive or even high measure (see Chapter 8).[13]

---

[11] Russell (1912, p. 21, our emphasis).

[12] For a critical discussion, see Callender (2004).

[13] Moreover, recall that entropy is measured by the v-measure, while probability is measured by the μ-measure, and there is no necessary connection between these two measures.

Probability is associated (in one way or another) with a lack of information concerning which, out of a set of possibilities, is the case.[14] Therefore, assertions about the probability of macrostates ought to be reserved for future states in situations where we do not know which states will obtain. It is meaningful to assign probability to *unknown* past states, *provided* that they imply possible events in the future, such as finding new records; but in this case the probability is actually about the future events, such as about finding those records. Whenever we feel that it is meaningful to assign probabilities to past events, this is presumably so just because these past events are correlated with as yet unknown records to be found in the future. But with respect to the known past, where we already know which states obtained, namely those that we remember, assigning probabilities to those states seems to us pointless, and it is not clear what purpose it could serve. And when the past is unknown, and has no implications for the future, not even in the sense of possible records, then assigning probabilities to past events is just as pointless. The only case in which we can meaningfully assign probabilities to events in the past is when these events lead to new predictions about new records to be found in the future.

Another way of looking at this is the following: consider the case in which t_{-1} is the time of the Big Bang or some other sense of beginning of the universe.[15] In that case, the notion of probability of the initial macrostate $[M_0]$ is not well defined since probabilities in statistical mechanics are *transition* probabilities, based on the Probability Rule. (We discussed the notion of a probability distribution over the initial macrostate, that is, the so-called Statistical Postulate, in Chapters 6 and 8. But note that the Statistical Postulate entails nothing about the probability of $[M_0]$ itself.) In another sense, the probability of $[M_0]$ in that case is strictly and trivially 1, since it is known to have been the actual state.

And so it seems that saying that the low-entropy macrostate in the past was *improbable* just because it has a small measure[16] is a highly misleading statement; and this, in a way, misses the whole point of assigning probabilities to macrostates in statistical mechanics.

---

[14] This is roughly true for all interpretations of probability.

[15] In classical mechanics there is no sense of beginning: trajectories are doubly infinite, and so the notion of a beginning of the universe has to come from some other theory. We assume here that this can be done without loss of coherence.

[16] See footnote 13.

10.8 The dynamics of the Past Hypothesis

The Past Hypothesis, with its components the Reliability Hypothesis and the low entropy assumption, has some interesting dynamical consequences, which we now describe.

Continuing the case described above, we assume that the present macrostate at time t_0 is $[M_1]$, and that we remember that in the past, at time t_{-1}, the macrostate was $[M_0]$, which had lower entropy. Consider now the same evolution, from the point of view that this observer had at time t_{-1}: at that time the observer observed the system, found it in $[M_0]$, and predicted its future, using the Probability Rule, based on the dynamical blob $B(t_0)$, as in Figure 10.8. (Suppose, for simplicity, that both the μ-measure of probability and the ν-measure of entropy are the Lebesgue measure.) Our observer predicts that entropy will increase, in accordance with the Law of Approach to Equilibrium. Since the Lebesgue measure of $[M_0]$ is smaller than the Lebesgue measure of $[M_1]$, the dynamical blob B (t_0) can cover only part of the future macrostate $[M_1]$.

But then, as the observer carries out a measurement at t_0, she finds that the actual macrostate is $[M_1]$. As we said above, this entails that *most* of the microstates in $[M_1]$ are non-historical. This fact raises the following difficulty. Our algorithm for prediction, based on the Probability Rule, says that an observer for which the present macrostate is $[M_1]$ should calculate the dynamical blob that starts in $[M_1]$ (and not in $[M_0]$). But why should such an algorithm be successful, given that we know that most of the microstates in $[M_1]$ are non-historical? The problem is this. Combine the cases described in Figure 10.8 (in which the observer carries out a prediction at time t_{-1}) with Figure 10.1 (in which the observer carries out a prediction at time t_0, after having found that at t_0 the actual macrostate is $[M_1]$). For all we know, mechanics is consistent with the scenario in which all the historical points that evolved from $[M_0]$ to $B(t_0)$ (as in Figure 10.8) evolved to only a part of $B(t_1)$, in the manner illustrated in Figure 10.10, and in this case the probabilities that the observer calculates at t_0, in order to predict the macrostate at t_1 (using the blob $B(t_1)$ as in Figure 10.1), would be refuted by experiment, since most of the blob $B(t_1)$ is empty and not going to materialize.

A solution for this problem can be given by the following dynamical hypothesis:[17] the historical points are always distributed *densely* over *all*

---

[17] This dynamical hypothesis may seem different from that proposed by Albert (2000, Ch. 4) because it is given in terms of dynamical blobs, which do not feature in Albert's approach. But essentially the two dynamical hypotheses are the same since blobs invariably originate in past macrostates.

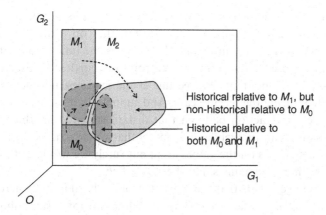

Figure 10.10 Why non-historical trajectories do not matter for prediction

the dynamical blobs. (We use "densely" and not "uniformly," since the latter notion depends on a choice of measure, whereas the former adapts itself to the measure μ of probability that is chosen according to observations.) This dynamical hypothesis explains why the algorithm based on the Probability Rule is successful.

10.9 Local and global Past Hypotheses

Recall that the Past Hypothesis, in Feynman's words, is " ... that in the past the universe was more ordered, in the technical sense, than it is today."[18] We now emphasize the word "*universe*" in this hypothesis. Does a postulation of a low-entropy initial macrostate *of the entire universe* explain the observed thermodynamic regularities, which hold for local subsystems of the universe, such as our paradigmatic ideal gas in a container? Suppose that one comes up with (a) a theory according to which the initial macrostate of the entire universe had low entropy;[19] and (b) a theorem according to which the entropy of the entire universe increases in accordance with the Law of Approach to Equilibrium. Would that entail that local subsystems also satisfy probabilistic versions of the Law of Approach to Equilibrium and the Second Law of Thermodynamics?

The answer is *no*, since the global evolution, of the dynamical blob of the entire universe, does not entail any kind of local evolution, in the

---

[18] Feynman (1965, p. 116).
[19] Again see Earman (2006) for a critical analysis of this idea in the context of cosmology.

subsystems of the universe that we experience, such as our paradigmatic gas in a container.[20] Suppose, for instance, that on a distant star the entropy increases significantly, and that at the same time (in a relativistic appropriate sense of the term) the entropy on Earth decreases significantly, say by a spontaneous freezing of the icebergs and glaciers that have melted during the recent years. And suppose that there is no significant physical interaction between that distant star and anything in the solar system. Quantitatively and formally that would satisfy the thermodynamic regularities on the global level, but surely we would not say that the laws of thermodynamics have been satisfied.

Until we have proofs that the properties of the local initial macrostates and of the *local* dynamics are such that the low entropy assumption holds for them, the only way to justify this assumption is by relying on our memories and records concerning these local systems, and *extrapolating* to similar cases.[21]

10.10 Past Hypothesis and physics of memory

The Past Hypothesis has implications for the way we understand the physics of memory states. By this we mean that the Past Hypothesis turns out to put some constraints on the kinds of physical processes that can bring about memories. Generally speaking, as expressed in the quotation from Russell above, memory is a state of affairs at present, which is presumably correlated with a state of affairs in the past. Clearly, the state of affairs at present is not of the systems featuring in the contents of the memory, but rather of the brain (say),[22] and the state of affairs in the past is also a state of the brain that – according to the Reliability Hypothesis – reflected the state of affairs of the world at that time. In Chapter 5 we saw that according to mechanics the physical state that *constitutes* the memory is a *microstate* of the brain. But the *content* of this memory is a feature of the *macrostate* of the universe (owing to the one-to-many correlation of the memory microstate with the rest of the world). The question is this: how does the present microstate of the brain bring about a reconstruction of the past microstate of the brain?

---

[20] For further discussion see Winsberg (2004), Earman (2006).
[21] Lanford's theorem is *not* a step in this direction since it suffers from the minimum problem; see Uffink and Valente (2010).
[22] Or of some other system, provided that the account is coherent with physicalism.

The first option (which we reject) is that a memory is the result of a *macroscopic retrodiction*, in which the observer calculates the past dynamical blob of the observed system, such as in Figure 10.4. That is, the brain processes realize such a calculation. A calculation of this kind would result in a probabilistic retrodiction, based on a time-reversed Probability Rule, and give rise to the minimum problem; our memories of the past, however, are not probabilistic. Even in cases where our memory is indecisive we say that we do not remember the past, not that we remember it probabilistically. As we said earlier, there seems to be no point in assigning probabilities to past events at all, except in cases where they give rise to future records.

And so the above option fails for two reasons: it is not only probabilistic but gives high probability to the wrong events. The second option (which we also reject) is that memory is based on some kind of *microscopic retrodiction*, in which the observer calculates the past microstate of its brain from the present microstate of its brain. But this option is implausible as well, for the following reasons. First, since the brain is a subsystem of the universe which, in general, interacts with other systems, retrodiction of the brain state must involve a retrodiction of all the systems with which it has significantly interacted during the relevant time interval, which together form an approximately isolated enough system. Second, even supposing that such a microscopic retrodiction were ideally possible, if our memory were a result of a microscopic retrodiction starting in the present brain microstate, we would remember every event that occurred in the past with exactly the same sharpness as we experienced them in the past. But, normally, we do not. Moreover, it would be hard to explain why we do not remember events that occurred in the remote past, for example in early childhood or even before we were born. And it would be hard to explain why we cannot remember the future, by carrying out similar microscopic calculations.

And so it appears that some elementary considerations concerning the mechanics of memory constrain the possible physical account of memory: it appears that memory is not a straightforward calculation of the past state from the present state. Of course, in the current state of physics it is extremely hard to see what would provide an understanding of the way in which certain microstates generate memory in particular, and mental experience in general. As we already remarked in Chapter 5, we do not profess to solve the mind–body problem.

10.11 Memory in a time-reversed universe

In order to further understand the Past Hypothesis and the notion of memory in it, it is instructive to consider this notion in a time-reversed universe. In Chapter 4 we saw that mechanics is compatible with a universe in which the direction of time is the opposite of the direction of time that we experience. In such a universe the arrow of time which forms part of each time-slice is opposite to ours, and therefore the time-slices are experienced in the reversed order. We showed that, in a sense, a film shown from the end to the beginning can illustrate the experience of an imaginary observer whom we called Tami, who lives in such a time-reversed universe. We are not interested in metaphysical speculations about such a universe *per se*, but we think that such a discussion can help us understand the implications of the principles of mechanics with respect to our actual universe. As we stressed in Chapter 4, realizing that Tami's universe is consistent with the principles of classical mechanics immediately implies that the laws of thermodynamics cannot be theorems of mechanics.

In this section we would like to address two claims, which sometimes appear in the literature, concerning Tami's possible memory states, as follows. (i) Tami has no mental states whatsoever, and so she has no memories in the usual sense;[23] Alternatively, (ii) if Tami does have memory states, then their contents are *her future* states, which are parallel to *our past* states.[24] Both claims are *wrong* in light of our discussions above, as we will now show.

Can Tami have mental states? It is extremely hard to say something constructive about how *our* mental states arise given what we know about our physical structure. At present, the best that we can hope for is to come up with some very general characterizations and constraints. For example, we assumed in Chapter 5 that our experience is associated with mechanical microstates, in a way that is consistent with identity physicalism (as well as with other theories of the mental); and that these microstates of the observer give rise to macrostates of the observed, through one-to-many physical correlations between them. Tami's microstates are identical to ours in all the details of all the time-slices, except the direction of time. Because of the great similarity between Tami's

---

[23] Maudlin (2005, Ch. 4).
[24] See, for example, Hawking (1988, Ch. 9) for this claim. However, Hawking attempts to underwrite the direction of our memories by the direction of entropy increase. In our view this line of reasoning is mistaken, but we shall not address this idea here.

time-slices and ours, we see no reason why Tami's brain microstates should not give rise to mental states.

On the other hand, owing to the difference between the time-slices, in their direction of time, the contents of Tami's mental states may very well be different from ours.[25] Since (at the present state of the art in all the sciences) we have no clue whatsoever as to how mental states are brought about by physical states, a significant difference between our and Tami's mental states, in parallel time-slices, cannot be ruled out. What can this difference be? Let us try to speculate, on the basis of what we know about statistical mechanics.

There seems to be no reason why Tami should not experience the world in terms of macrostates, and even of thermodynamic macrostates, since the physical correlations between microstates are assumed to be the same in both universes. Moreover, if the macrostates are the same and the dynamics is the same, the probabilities should be the same: Tami's predictions will be parallel to our retrodictions, illustrated in Figure 10.4. But whereas for us our memories select the actual past macrostates (as in Figures 10.5 and 10.6), for Tami this direction is the unknown future. Indeed, given that the microstates of Tami's brain are different from ours in the direction of time, they can give rise to different contents from ours, and in particular there is no reason why the contents of Tami's microstates should not reflect events that are in her past (which is our future). For Tami, Figures 10.5 and 10.6 (which for us represent the working of memory) would represent detection in measurement, whereas Figures 10.1, 10.2, and 10.3 (which for us represent a measurement) would represent memory. This gives further emphasis to the parallel structure of measurement and memory.

Thus, for every process in which we experience entropy increase, Tami will experience entropy decrease. If thermodynamics were universally true in our universe, Tami would have to have a universally true *anti*-thermodynamic theory. Assuming, therefore, that mechanics is true in both universes, the laws of thermodynamics and their probabilistic counterparts in statistical mechanics cannot be theorems of mechanics.

---

[25] Natural Selection cannot be brought in to account for this difference in a physicalist framework, precisely because the time-slices are exactly identical in everything but the direction of time.

11
Gibbs

11.1 Introduction

In the standard literature[1] on the foundations of statistical mechanics, one is presented with two theories, both of which are called statistical mechanics: one originates in the work of Josiah Willard Gibbs[2] and the other in the work of Boltzmann. Many of the recent accounts of the philosophical foundations of statistical mechanics prefer the tradition set by Boltzmann, for good reasons pertaining to the conceptual basis of the two theories.[3] However, the empirical predictions based on the Gibbsian approach are successful in a wide range of phenomena, and are prevalent in standard physics textbooks on statistical mechanics; and this fact calls for an explanation. In this chapter we wish to present some of the ideas of the Gibbsian approach in terms of our conceptual framework and explain its empirical success.

Up to now our reconstruction of statistical mechanics, and our understanding of the way in which probability arises in the deterministic framework of classical mechanics, has been based on the interplay between macrostates and dynamical blobs. Although calculating dynamical blobs is extremely difficult and may not be feasible, we saw in Chapter 7 that there are shortcuts that enable us to avoid these calculations, owing to some contingent special characteristics of the actual dynamics of the universe. But this still leaves us with another major obstacle: a precise identification of the regions in the state space that correspond to thermodynamic *macrostates* is also extremely difficult. This

---

[1] See Sklar (1993), Uffink (2007), Frigg (2008).
[2] Gibbs (1902); see also Tolman (1938), Guttmann (1999).
[3] See Lebowitz (1993), Callender (1999), Goldstein (2001), Albert (2000).

was our very important remark from Section 5.2. The partitioning to macrostates is determined by the structure of the accessible region and the details of the interactions between the observer and the observed, and these details are often beyond our knowledge. For this reason, we need another shortcut that will allow us to express the thermodynamic regularities without appealing to the precise structure of the macrostates. And it turns out that there is a standard approach to statistical mechanics, one that can be understood as providing such a shortcut, at least for some important thermodynamic cases: the method proposed by Gibbs. Let us now present the general outline of Gibbs's ideas and see how they provide this shortcut.

11.2 The Gibbsian method in equilibrium

Consider a thermodynamic system (such as our paradigmatic example of an ideal gas in a container) in a state of thermodynamic equilibrium. In thermodynamics we characterize equilibrium by the empirical fact that the measured values of thermodynamic magnitudes are stable over time. In mechanics, by contrast, we say that the thermodynamic magnitudes may fluctuate, but in equilibrium (as defined in Chapter 7) the fluctuations are so small, fast, and unpredictable that we can hardly detect them and cannot exploit them to produce work, and so for all practical purposes they are negligible.

We could, perhaps, exploit the fluctuations if we could detect them with our measuring devices; but such detection is practically impossible, for the following reason. Thermodynamic systems, such as our paradigmatic gas (call it G), evolve fast relative to measuring devices and observers (call them D): the microstate of D depends on the microstate of G, but it takes time for D to change its microstate so that it will reflect the microstate of G, and by that time the microstate of G has already undergone considerable change. This means that although the microstate of D depends on the microstate of G, the microstate of D, as it were, lags behind the microstate of G, and reflects its past microstates. Moreover, this reflection is not one-to-one, since by the time D has reacted to some past microstate of G, it has already been subjected to the effects of subsequent microstates of G as well. The result is that the measurement interaction has induced some correlations between the final microstate of the measuring device D and some general features of the evolution of the measured system G, such as some *time average* over the microstates that

G's trajectory has gone through during the interaction. And since G's microstates during the measurement interaction can in general belong to different macrostates of G, D's final state does not even reflect a specific macrostate of G, but rather some *time average over the macrostates* that G passed through during the interaction.

By assumption, the measured system G is in thermodynamic equilibrium: that is, its thermodynamic magnitudes are found to be stable upon repeated measurements by the measuring device D. This means that the averages over the microstates of G reflected in the microstate of D are the same for all the measurements. And to say that G is in equilibrium is to conjecture that subsequent measurements of G by D will reveal the same average. This conjecture is dynamical: it reflects a hypothesis concerning the dynamics of G and D, which we now describe.

The Gibbsian dynamical hypothesis says that during the time it takes for D to reflect a microstate of G, G's trajectory has enough time to travel through regions in the state space that are representative of the entire accessible region. By *"representative"* we mean that if we take state averages along segments of the trajectory of G that have finite *lengths* which correspond to the thermodynamic relaxation *time*, these state averages are equal to averages taken over the entire accessible *region*, i.e. over all the possible infinite trajectories of G. This may not be true for all the trajectory segments, but we assume that it is true for *most* segments that have the minimum length corresponding to the thermodynamic relaxation time. That is, we assume that most (by some measure) of the points in the accessible region are initial points of finite trajectory segments that have the same state average.[4] One attempt at constructing a foundation for Gibbs's approach along these lines is to assume that the system's dynamics is ergodic,[5] and to add (often implicitly) the assumption (which does not follow from ergodicity) that the finite segments that are observed are representative of the entire accessible region.[6]

In order to make the dynamical hypothesis precise, one has to add some details concerning the way in which averages should be calculated: because an average over the continuum of microstates in a trajectory or

---

[4] Instead of considering microstates that are the initial states of trajectory segments, Gibbs preferred the pictorial notion of an *ensemble*, which is an infinite collection of identical systems, each of which starts out in a different microstate. See Gibbs (1902, p. v). For the notion of an ensemble, see Peres (1993, p. 25, fn. 1, p. 59, fn. 9).

[5] Recall from Section 3.7 that the main feature of ergodic systems is that along ergodic trajectories the time average is equal to the state-space average.

[6] For an overview of problems in this attempt, see Sklar (1993), Earman and Redei (1996), and Frigg (2008).

over regions in the state space is relative to a *measure*. For the finite averages to be constant over time, namely to be adequate for the state of equilibrium, Gibbs proposed to take probability measures that are stationary under the dynamics. Obviously, Gibbs's proposal delivers the goods provided that the dynamical hypothesis is satisfied. And so, the combination of the dynamical hypothesis with stationary probability measures guarantees a notion of equilibrium in terms of state-space averages.

However, stationary probability measures are not enough if we are to recover the empirically correct thermodynamic predictions. The mere identification (or postulation) of such measures does not suffice to generate, for example, the thermodynamic equilibrium state equations, such as the ideal gas law. In Gibbs's method, one must also identify the right sorts of averages that ought to be taken relative to these measures. Gibbs's great achievement was to associate with each set of thermodynamic constraints a stationary probability measure, and, for each such measure, to identify various averages that correspond to the thermodynamic magnitudes under these constraints. Gibbs calculated two kinds of such quantities: one is *state averages* over the entire accessible region, based on a given measure; and the other is *functions of the measure* itself, again over the entire accessible region. Gibbs showed that these quantities, relative to each of the stationary measures, have functional relations between them that match the functional relations between thermodynamic magnitudes in equilibrium, for the corresponding set of constraints. Because of this similarity in functional role, he proposed to consider these quantities as *analogous* to the thermodynamic magnitudes under the suitable external constraints.

For example, in a system with fixed energy[7] both entropy and temperature are functions of the probability distribution that is uniform relative to the Lebesgue measure (or volume) over the entire accessible region in the energy hypersurface of the system. That is, entropy is given by $\log \Omega(E)$ and temperature by $\left(\frac{d\log \Omega(E)}{dE}\right)^{-1}$, where $\Omega(E)$ is the Lebesgue measure of the accessible region. In a system with constant temperature,[8] temperature is given by a parameter θ that determines the probability distribution according to which the microstates with energy in the interval dE have probability $p = e^{E/\theta}dE$. Entropy in this case is given by $\int p \log p$ over

---

[7] This is the so-called *microcanonical* ensemble.
[8] This is the so-called *canonical* ensemble.

the entire accessible region.[9] In the scenario of measurement described above, one can say that the microstate of the thermometer, for example, is determined by the parameter θ rather than by any other feature of the trajectory of *G*. Note that these quantities may be interpreted also as state-space averages (relative to the appropriate measures).[10]

Later physicists took this idea a step further and *identified* the Gibbsian quantities with the thermodynamic magnitudes.[11] This identification entails that the chosen measures have to be such that the quantities based on them yield empirically adequate predictions, not only with respect to functional relations, but also with respect to each of the quantities on its own. (We emphasize that the Gibbsian analogies do not necessarily entail a unique choice of measure.)

In sum, we can interpret the Gibbsian method for underwriting equilibrium thermodynamics in a way that yields predictions concerning individual systems as follows. First, put forward the dynamical hypothesis. Second, find stationary probability measures such that if these measures are used to calculate state-space functions and averages, then given the dynamical hypothesis, these quantities will satisfy the thermodynamic functional relations under various sets of constraints. Third, identify each probability measure with the equilibrium state given a suitable set of thermodynamic constraints; and identify the various averages and functions obtained from these measures with the thermodynamic magnitudes under these constraints. Once all this is obtained, these measures and averages can be used to produce further magnitudes that can form a basis for predictions.

11.3 Gibbsian method in terms of blobs and macrostates

We now turn to describe this Gibbsian method in terms of the conceptual framework of blobs and macrostates. In particular, we wish to show that our notion of equilibrium – as a macrostate with certain features – is related to Gibbs's notion of equilibrium as a stationary probability measure. Recall (Chapter 7) that the macrostate of equilibrium $[M_{eq}]$ is

---

[9] For details of these examples, see Uffink (2007, Sec. 5.1).

[10] For a comparison of the Gibbsian notion of entropy with the Botzmannian one, see Jaynes (1965).

[11] See, for example, Peres (1993, Ch. 9. p. 270). This is a more reductionist approach than that of Gibbs, who tried to avoid reductionism, since he suspected that some problems in physics required a significant change in the fundamental theories.

characterized such that the *transition* probability (given by the suitable probability measure μ) of evolving away from $[M_{eq}]$ to any other macrostate is small. This means also that the μ-measure of $[M_{eq}]$ is large. Now if we take an average relative to the μ-measure over the entire accessible region, we obtain approximately the same μ average taken over $[M_{eq}]$ only. Since all the microstates in $[M_{eq}]$ are indistinguishable from one another (by the observer O) this means that the Gibbsian average taken over the entire accessible region characterizes each and every microstate in $[M_{eq}]$, provided that the measure in the Gibbsian average is the μ-measure. Moreover, since the μ averages over finite segments are equal to the μ averages over the entire accessible region (by the dynamical hypothesis) it turns out that the μ averages over finite segments are approximately equal to the value of any individual microstate in $[M_{eq}]$.

Recall that in our approach, the probability measure μ is chosen so as to fit the relative frequencies that we see in experience (and chosen so that it is convenient). In order to be successful, the Gibbsian method has to use the same measure. It follows that one can justify the choice of the probability measures in the Gibbsian method in the same way, namely, on the basis of empirical generalization and convenience. (We emphasize this point since some writers tend to think of the Gibbsian probability distributions as *a priori* to some extent.)[12]

An instructive similarity between the Gibbsian method and the one proposed in this book concerns a difference between the two kinds of quantities that give rise to the Gibbsian thermodynamic analogs. As we said, some of these quantities are state averages, which have units that correspond to the mechanical states; and other quantities are functions of the measures themselves, and have no physical units. Entropy belongs to the latter kind: for example, the entropy of a system with constant energy[13] is the logarithm of the measure of the accessible region, and the entropy of a system with constant temperature is given by $\int p \log p$, where p is the measure and the integration is taken over the entire accessible region. This difference is paralleled in the approach we put forward in this book, where entropy is a function of the measure of a macrostate, while the mechanical counterparts of other thermodynamic magnitudes are functions of the microstates and therefore have suitable units. (Incidentally, the fact that entropy is a function of the probability measure enables the use of this quantity in domains remote from

---

[12] This tendency is best seen in the work of Jaynes (1957, 1965).
[13] In terms of Gibbs's ensembles this is the microcanonical ensemble.

thermodynamics and statistical mechanics; unfortunately, this conceptual expansion sometimes leads to confusion.)[14]

This Gibbsian treatment enables us to recover important aspects of the thermodynamic regularities in equilibrium, *without the need to know the detailed partitioning of the state space to macrostates*, in the case where (i) the dynamics is such that the Law of Approach to Equilibrium is satisfied; and (ii) the Gibbsian averages are taken using the μ-measure used in the Probability Rule, and chosen so as to fit observed relative frequencies and be convenient. Wherever applicable, this is an important and very useful shortcut. In fact, this is textbook statistical mechanics as we know it.

11.4 Gibbsian equilibrium probability distributions

However, one important aspect of equilibrium is missing in the Gibbsian method for equilibrium, namely the notion of *fluctuations*. The only role of probability measures in this method is in calculating averages, and these averages are taken to generate *deterministic* empirical predictions. They are not understood as generating probabilistic predictions, and hence do not generate a prediction of fluctuations. In this sense, the Gibbsian method is not probabilistic.

Indeed, up to now we have explained the success of the predictions based on the Gibbsian method on the assumption that whenever a measurement is carried out on G we hit upon a trajectory segment which is representative; and so the average along it is equal to the average over the entire accessible region. However, in order to explain why in experience we hit upon representative segments, we need to add the assumption that representative segments have high probability. Given this assumption, it is practically certain that our measurements will yield the outcomes corresponding to the calculated average. But since not all the trajectory segments are representative, and since any trajectory may contain non-representative segments, it is not impossible that our measurements will

---

[14] Claude Shannon recalls that, in the development of the physics of communication channels, he sought for a name to one of the central quantities: "My greatest concern was what to call it. I thought of calling it 'information', but the word was overly used, so I decided to call it 'uncertainty'. When I discussed it with John von Neumann, he had a better idea. Von Neumann told me, 'You should call it entropy, for two reasons. In the first place your uncertainty function has been used in statistical mechanics under that name, so it already has a name. In the second place, and more important, nobody knows what entropy really is, so in a debate you will always have the advantage.'" (Tribus and McIrvine, 1971).

hit upon such a segment, in which case the outcomes we observed would correspond to a *fluctuation*. In experience large fluctuations away from thermodynamic equilibrium are never seen, but very small fluctuations are not that rare: they are empirically observable, and are not impossible by the underlying mechanics. How can this empirical fact be accounted for in the Gibbsian conceptual framework?

Intuitively, two options come to mind. One can either say that a system fluctuates *out-of*-equilibrium, or that it fluctuates *in* equilibrium. However, the trouble is that neither of these two options is well defined in the Gibbsian framework. Consider the first option. In the Gibbsian framework, equilibrium is defined by stationary probability measure. But the fluctuations we are talking about are statistical consequences *given* the measure that defines the equilibrium. Therefore in the Gibbsian approach it makes no sense to suggest that they are fluctuations away from equilibrium. On the other hand we cannot say with the second option that the fluctuations are in equilibrium since it can be empirically verified that the measured thermodynamic values are no longer those of equilibrium. A way to solve this difficulty is by biting the bullet and saying that each fluctuation involves a change of the probability measure. We will see that a similar strategy can be applied also in the Gibbsian account of the approach to equilibrium.

11.5 The approach to equilibrium

Consider our usual example of a gas expanding in a container, illustrated in Figure 1.1. At time t_0 a gas is confined by a partition to the left half of a container. At time t_1 the partition is removed. As we know from experience the gas will evolve gradually and spontaneously through an intermediate stage at t_2 to fill up the entire volume of the container at t_3. This evolution satisfies both the Law of Approach to Equilibrium and the Second Law of Thermodynamics. As we saw in Chapter 2, the thermodynamic Law of Approach to Equilibrium describes evolutions in cases where the external constraints remain unchanged, and the final equilibrium state is uniquely determined by these constraints. The case of changed constraints is covered by the Second Law of Thermodynamics, which asserts that upon a change of constraints, once a system has evolved from the equilibrium corresponding to the first set of constraints to the equilibrium corresponding to the second set of constraints, its final entropy is not lower than its initial entropy.

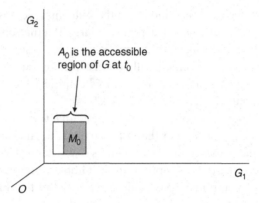

Figure 11.1 State-space description of a gas confined to the left-hand side of a container

In the Gibbsian way of accounting for equilibrium, as presented above, the counterpart of the Second Law should be that the Gibbsian entropy at t_3 (which is a function of the probability distribution in the equilibrium state corresponding to the final constraints) cannot be smaller than the Gibbsian entropy at t_0 (which is a function of the probability distribution in the equilibrium state corresponding to the initial constraints). This Gibbsian formulation of the Second Law is not a theorem of mechanics, and its truth depends on assumptions concerning the dynamics that governs the actual evolution of the gas. This evolution, in turn, may or may not satisfy the Law of Approach to Equilibrium; and it presents conceptual problems for the Gibbsian approach, of which Gibbs seems to have been aware. He proposed the so-called coarse-grained approach to equilibrium which, for finite times, violates Liouville's theorem, as we shall see. We will now propose a way of reading Gibbs's proposal in our framework which may solve this problem.

Consider the state-space description of the evolution of our gas expanding in a container illustrated in Figures 11.1 and 11.2 (where G_1 and G_2 are two representative degrees of freedom of G).

At t_0 the accessible region is determined by the constraint of the partition: the trajectories are confined to the region corresponding to positions on the left-hand side of the container. At t_1, when the partition is removed, the accessible region, and with it the structure of the trajectories, change instantly, such that all the positions in the container become accessible for the gas molecules, and this new structure of the trajectories remains unchanged throughout the rest of the experiment. But of course, at t_1 the system is still not in equilibrium, and therefore it would be wrong

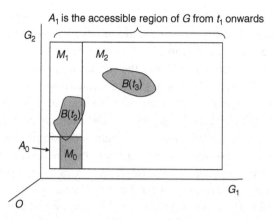

Figure 11.2 State-space description of a gas expanding in a container

to say that the probability distribution at t_1 is already the stationary one determined by the constraints. Measurements at t_1, t_2, and t_3 are expected to yield different outcomes, which means that the average microstate of G during these measurements, which is reflected in the measuring device D, changes between t_1 and t_3. And since the average is taken over the entire accessible region, the change can come (in the Gibbsian method) only from a change in the probability distribution. Indeed, for example, immediately after t_1 the probability of the gas G being in a microstate in which it has expanded in the whole container is zero, while after t_3 this probability is high. Thus we can say that the evolution of the probability distribution supervenes on the dynamics.

How can one express such a gradual dynamical change in the probability distribution in the Gibbsian framework? Gibbs offers a general qualitative discussion of this point, known as the approach to coarse-grained equilibrium; this seems to assume a conceptual framework that is hard to reconcile with Gibbs's own framework of the account of equilibrium. Here, however, we do not attempt to present Gibbs's own ideas. Rather, we want to show how some of Gibbs's insights can be understood as a practical shortcut for applying our conceptual framework in a way that precludes the need to identify the exact regions that correspond to macrostates. With this in mind, here are the conditions in which Gibbs's method can be used, and in which applying this method will be successful.

First, at t_0, if the measure of the equilibrium macrostate covers almost the entire accessible region, such that $\mu([M_0]) \cong \mu(A_0)$ (see Figure 1.1), then the average, using the μ-measure, over the entire A_0, will be approximately equal to an average, using the μ-measure, over $[M_0]$ only; and this, in turn, will reflect the state of the system G which the observer O experiences when G is in any of the microstates in $[M_0]$. This is an account of the success of the Gibbsian method in the equilibrium state $[M_0]$ at t_0. The same holds at t_3, when again G is in the equilibrium macrostate, which is now $[M_2]$ (see Figure 11.2), where the accessible region is A_1.

Gibbs now encounters a problem. On the one hand, the Second Law asserts that the entropy of the gas at t_3 in our example is higher than the entropy at t_0. On the other hand, the dynamical blob at t_0 is exactly the same as the blob at t_1, and from t_1 to t_3 the blob $B(t)$ *evolves* subject to Liouville's theorem. This means (in those cases where the Gibbsian entropy is given by a function of the Lebesgue measure) that the entropy of the gas during the evolution from t_1 to t_3 cannot increase (or decrease). To address this problem Gibbs proposed the so-called coarse-grained approach to equilibrium, which goes roughly as follows.

The dynamical evolution of the blob $B(t)$ is such that with time the blob *blends* with the rest of the accessible region. In more detail, Gibbs postulates that the dynamical blob becomes fibrillated so that as time goes to infinity the points of $B(t)$ become dense amongst the points that do not belong to the blob. That is, every small region of the state space contains both points that belong to the blob and points that do not belong to the blob. Gibbs takes this to mean that the observer can no longer tell whether any given small region at a given moment is within the dynamical blob, and therefore this observer begins to treat the entire accessible region as possible at all times after t_3. Gibbs concludes that the entropy of the gas now calculated over the entire accessible region has increased. But this conclusion violates Liouville's theorem.[15]

The way to solve this problem is by introducing a clear distinction between the dynamical blob, which is subject to Liouville's theorem, and macrostates which determine the entropy and which are not subject to the theorem. In his coarse-graining argument Gibbs introduced the idea of an evolution of a dynamical blob, but did not develop the further step of realizing the interplay between blobs and macrostates in the case of an evolution *towards* equilibrium. A way to solve this problem is by returning to the idea that we already encountered in the case of equilibrium (at the very end of Section 11.4). One can think of the approach to equilibrium dynamically in the Gibbsian picture as a sequence of changing probability distributions (or measures) along which initially at

---

[15] For a critical analysis of this idea, see Ridderbos and Redhead (1998) and Ridderbos (2002).

t_1 the measure defined over the entire accessible region is concentrated in the region $[M_0]$, at t_2 the measure is mostly concentrated in the region $[M_1]$, and from t_3 onwards the measure concentrates mostly in the region $[M_2]$, which is the Gibbsian equilibrium measure. Conceptually this may solve the problem of the coarse-grained entropy in the Gibbsian method. However, in contrast to the case of equilibrium where the Gibbsian method provides a feasible way of avoiding hard calculations, it seems, in this case, that the calculations that it involves are insurmountable.

12

Erasure

12.1 Introduction

The notion of *information*, which is central in classical statistical mechanics, is objective and physical. As we emphasized in Chapter 9, the notion of information is part of the physical definition of macrostates: on the one hand, the observer in question has no information about which microstate in a given set is the actual microstate at each moment; on the other, the observer does have information about which is the macrostate that contains the actual microstate. Both the information concerning which is the actual macrostate, and the lack of information concerning which, in this macrostate, is the actual microstate, are objective features of the physical correlations between the observer and the observed.

The information concerning the actual macrostate is obtained in measurement, which is a special kind of dynamical evolution given in physical terms. Now, once the observer acquires information concerning which is the actual state of an observed system, this information can be manipulated. One important kind of manipulation is *erasure*, which is a physical operation after which the observer no longer has information concerning the pre-erased state of the system; in other words, the microstate of the observer is no longer correlated with the pre-erased macrostate of the observed system. Although erasure has special importance, there are other kinds of manipulations of information, that is, other kinds of physical implementation of logical operations, some of which will be presented in this chapter.

The conventional wisdom in the literature on the physical implementation of manipulation of information is that logically irreversible operations such as erasure *necessarily* involve entropy increase by $k \log 2$ per bit of erased information. This idea was proposed by Rolf

246

Landauer,[1] and is currently prevalent and predominant.[2] Landauer's thesis and the main argument for it are best summarized in Landauer's own words:

> Consider a typical logical process, which discards information, e.g., a logical variable that is reset to 0, regardless of its initial state ... The erasure process we are considering must map the 1 space down into the 0 space. Now, in a closed conservative system phase space cannot be compressed, hence the reduction in the spread [in the degrees of freedom representing 1 and 0] must be compensated by a phase space expansion [in other degrees of freedom], i.e., a heating of the irrelevant degrees of freedom, typically thermal lattice vibrations. Indeed, we are involved here in a process which is similar to adiabatic magnetization (i.e., the inverse of adiabatic demagnetization), and we can expect the same entropy increase to be passed to the thermal background as in adiabatic magnetization, i.e., $k \ln 2$ per erasure process. At this point, it becomes worthwhile to be a little more detailed ... This is, however, rather like the isothermal compression of a gas in a cylinder into half its original volume. The entropy of the gas has been reduced and the surroundings have been heated, but the process is not irreversible: the gas can subsequently be expanded again. Similarly, as long as 1 and 0 occupy distinct phase space regions, the mapping is reversible. The real irreversibility comes from the fact that the 1 and 0 spaces will subsequently be treated alike and will eventually *diffuse* into each other.[3]

In this chapter we will explain this thesis and show that while it may hold in interesting and even familiar situations, it does not logically follow from the principles of classical mechanics. This means that there is no special thermodynamic significance to the logical properties of the logical operations that are implemented in physical systems. The thermodynamic properties of such physical processes depend only on the physical details of these processes. In addition we shall analyze and generalize Landauer's idea of diffusion and put forward the necessary and sufficient conditions for erasure.

12.2 Why there is no microscopic erasure

In Chapter 9 we said that the notion of measurement cannot be defined at the level of microstates, since on that level there is no ignorance, because

---

[1] Landauer (1961).
[2] See Lef and Rex (2003). A version of this thesis has been proposed also by Penrose (1970). For a critical presentation of the Gibbsian version of these theses, see Maroney (2005).
[3] Landauer (1992, p. 2), our emphasis.

of determinism. As we will see in this chapter, a similar kind of reasoning leads to the conclusion that the physical implementation of logically irreversible operations are also macroscopic and have no *micro*scopic counterpart. Let us see why.

In a trivial sense, the microstate of the universe at present is a memory of all the past microstates of the universe, since given the equation of motion one can derive any past (as well as future) microstate from the present microstate. Therefore microscopic memory can never be erased.

Another way of looking at this matter is this. In the context of the physical implementation of computation (namely, in computers), an erasure is a physical implementation of the function $f(0) = f(1) = 1$ or equivalently $f(0) = f(1) = X$ (where X is a standard state). These functions are erasures since from the final state one cannot infer the initial state. This kind of erasure cannot be implemented at the microscopic level in the case of classical mechanics because of the determinism of the dynamics: determinism means that two different microstates, such as those implementing the data 0 and 1, cannot both evolve to the same microstate, such as 0 or 1.

And so it turns out that erasure (as well as other logically irreversible operations) can be carried out only on macrostates.

12.3 What is a macroscopic erasure?

Erasure is the reversal of measurement, in the sense that it begins in a (post-measurement) state in which an observer knows which of two states actually obtains, and ends in a (pre-measurement) state in which the observer does not know which of the two cases obtained. However, this transformation involves some subtle details that differ from those of measurement, and we now turn to describe them.

Consider the measurement, described in Figures 12.1, 12.2, and 12.3. Figure 12.1 describes the pre-measurement macrostate $[M_0]$, and each of the other two figures describes one of the possible outcomes of the measurement, $[M_1]$ or $[M_2]$, after the split and collapse stages took place (to recall the meaning of these terms see Chapter 9). At this final stage, by looking at the macrostate of D the observer O can infer the macrostate of G.

Erasing this measurement's outcome means bringing about an evolution such that O will no longer be able to infer the post-measurement state of $D + G$ by looking at D. To achieve this, the erasure must bring about a state that will be compatible with both post-measurement (or pre-erasure)

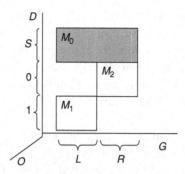

Figure 12.1 Post-erasure macrostate

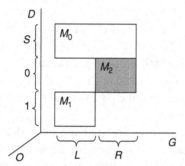

Figure 12.2 Pre-erasure macrostate: part a

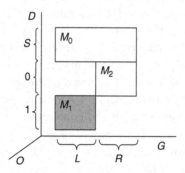

Figure 12.3 Pre-erasure macrostate: part b

outcomes, and thus it will be impossible to infer from that final state which outcome obtained. The simplest way to do this is by carrying out the reversal of the measurement: to map both the trajectories starting out in $[M_1]$ and the trajectories starting out in $[M_2]$ to the same macrostate, in this case back to $[M_0]$. Figures 12.2 and 12.3 are now the *pre-erasure* states, and Figure 12.1 is the *post-erasure* state. This mapping is an erasure because, by assumption, the measuring device D cannot distinguish between the subregions of $[M_0]$ that came from $[M_1]$ and those that came from $[M_2]$, and hence this information is lost to O as well. (We will later discuss the case where D can distinguish between these two subregions of $[M_0]$.)

Sometimes in the literature one encounters a description of erasure that looks like a superimposition of all or two of the above three Figures 12.1, 12.2, and 12.3. This should be understood in one of the following two senses, depending on the context: (1) a shorthand version of the description given by the above different figures; (2) a description from the point of view of an observer who prepares $D + G$ in the macrostate $[M_0]$, and predicts the outcome of a measurement followed by an erasure. In this case, the shading of $[M_0]$ is indeed a description of the *initial* macrostate, the shading of $[M_1]$ and $[M_2]$ should be understood as the dynamical blob (and not actual or counterfactual macrostates) after the measurement and before the erasure, and then the $[M_0]$ is also a representation of the final post-erasure dynamical blob.

In the special case that we are studying here, the measurement maps one macrostate to two, and the erasure maps two macrostates to one. While this is a very special case (and a more general one will be presented below), we can already learn from it an important aspect of the connection between erasure and its corresponding measurement. In the measurement the initial macrostate $[M_0]$ contains microstates which are not going to be realized in the future, even macroscopically, since they are going to evolve into the macrostate which is not going to be the outcome of the measurement. That is, as we said in Chapter 9, the initial blob splits in the course of the evolution, so that part of it overlaps with the macrostate which turns out to be the actual outcome of the measurement; and another part of it overlaps with the counterfactual macrostate, which does not materialize. This latter part is ignored after the measurement, since the observer knows with certainty that it does not contain the actual microstate. Similarly, in the corresponding erasure, the end macrostate $[M_0]$ must contain *non-historical* points, namely points which arrive from the macrostate of the counterfactual measurement outcome, that is the outcome that did not actually occur.

For simplicity, we referred in the above figures to the special case in which the Lebesgue measure of the union of $[M_1] + [M_2]$ is equal to the Lebesgue measure of $[M_0]$. In this special case the Lebesgue measure is trivially doubled in the erasure, so that entropy (if calculated by this measure) increases by $k \log 2$, which is the amount by which entropy has decreased in the corresponding measurement (see Chapter 9). This case appears to correspond to Landauer's thesis.

However, the set up described in Figures 12.1, 12.2, and 12.3 is a special case, and may give the wrong impression that there is an intrinsic linkage between erasure and entropy increase. Let us now consider a more general set up from which we can learn that there is no such link, just as there is no link between measurement and entropy increase or decrease (as we saw in Chapter 9). Using the more general case we will also describe the necessary and sufficient conditions for erasure.

12.4 Necessary and sufficient conditions for erasure

A necessary and sufficient condition for an erasure is what we call *blending*, which is similar to Landauer's notion of *diffusion*. Landauer,[4] in the quotation cited at the introduction to this chapter, says that the completion of an erasure requires what he calls *diffusion*. The set up that Landauer seems to have had in mind is of the kind sketched in Figures 12.1, 12.2, and 12.3 where, after the two bundles of trajectories arrive at $[M_0]$ from $[M_1]$ and $[M_2]$, they diffuse into each other, and are no longer distinguishable. Landauer may have thought that after some time any region of positive Lebesgue measure within $[M_0]$ contains both end points that came from $[M_1]$ and end points that came from $[M_2]$, and so it is impossible to identify subregions in $[M_0]$ that contain end points that belong to only one of these bundles. The idea of diffusion is indeed very important, but it can and should be generalized; and the generalization entails – as we will now see – that Landauer's thesis concerning the entropy of erasure is not a theorem of mechanics.

To see this, consider first Figures 12.4 and 12.5, which illustrate the necessary and sufficient condition for erasure that we call *blending*. Recall from Chapter 9 that the split condition implies that the dynamical blob evolved from the initial pre-measurement macrostate to the *two* macrostates corresponding to the measurement outcomes, in such a way that it partly overlapped with these two macrostates, while satisfying Liouville's theorem. Alternatively, in a very special case, the blob could fully overlap

---

[4] Landauer (1992).

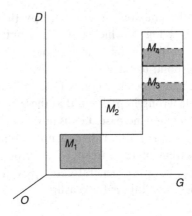

Figure 12.4 Blending: part a

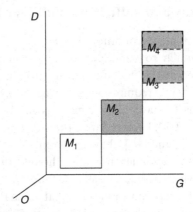

Figure 12.5 Blending: part b

with the two final macrostates. In Figure 12.4 the idea is similar: all four macrostates have the same Lebesgue measure; and the trajectories that start in the macrostate [M_1] evolve in such a way that the dynamical blob partly overlaps with macrostates [M_3] and [M_4]. In this special case, designed for simplicity, the blob overlaps with exactly ½ of [M_3] and ½ of [M_4]. Similarly, in Figure 12.5 we see the evolution of the trajectories that start in [M_2]: they also evolve so that their dynamical blob overlaps with exactly the remaining ½ of [M_3] and the remaining ½ of [M_4].

By the end of this evolution, O measures the state of D in order to learn from it the state of G. Here, the usual measurement process takes place (see Chapter 9). Assuming that O is correlated with D and G in such a way that O is able to distinguish between [M_3] and [M_4], a detection takes place and then the dynamical blob collapses on either [M_3] or [M_4], depending on which of them contains that actual microstate of $D + G$. By the end of this measurement, O can say only that the macrostate of $D + G$ is the outcome of the measurement, but cannot tell which subregion of the actual macrostate contains the actual microstate, for this is the very idea of the notion of a macrostate. And so, if the blobs that started in [M_1] and [M_2] blend within [M_3] and [M_4], by looking at D the observer O cannot tell whether the original macrostate was [M_1] or [M_2].

The case in which the correlations between O and $D + G$ are such that O can distinguish between the macrostates [M_3] and [M_4] is special: in general this correlation can be either finer or coarser. An example of an erasure with a finer correlation is addressed in the next section and in Figure 12.6; the case of an erasure with a coarser correlation is illustrated above, in Figure 12.1. It is very important to realize that there is *no* intrinsic connection between blending (or diffusing), which depends on the structure of trajectories in the blobs, and/or the entropy, which is fixed by the measure of the macrostates. We have already seen blending dynamics that increases the entropy, such as the one in Figure 12.1, and blending dynamics that conserves entropy, such as the one in Figures 12.4 and 12.5. In the former case the blending takes place *within* a single macrostate, and in the latter case the blending (or diffusion) occurs *across* different macrostates.

It is crucial to see that erasure ends in a macrostate that depends on the *contingent* correlation between O and $D + G$. To see this point, consider the case in which after the erasure, the observer does not carry out the measurement that would distinguish [M_3] from [M_4]. In this case, by the end of the erasure the microstate of O is correlated with the union of [M_3] and [M_4] (this situation is essentially the same as the case depicted in Figure 12.1), and this union is the final macrostate. Although O did not carry out an active process, which we would normally call "measurement," O has significant information concerning D and G, owing to the fact that O knows that an erasure has taken place on the specific systems D and G. This information is a physical correlation and, in physical terms, is no different from the active process we normally call "measurement." And so, the final macrostate – and *ipso facto* the final entropy – of an erasure, depends on the matter of fact concerning the contingent correlations between O and $D + G$.

In familiar thermodynamic situations there seem to be fixed limitations on the observation capabilities of human observers and, in this sense, one can perhaps introduce a maximally fine-grained partition to thermodynamic

macrostates, which results in some specific entropy of post-erasure macrostates. But the details of these macrostates are a contingent matter of fact. In
particular, the principles of mechanics entail no specific relation between the
pre-erasure and post-erasure entropy of the universe. This means that
whether or not Landauer's thesis is true for the familiar thermodynamic
situations is a question of contingent fact as well. In any case, contrary to the
conventional wisdom, *Landauer's thesis is not a theorem in classical
mechanics.*[5]

This completes our account of the notion of blending, which is both
necessary and sufficient for erasure in classical mechanics.

12.5 Logic and entropy

In the special case illustrated in Figures 12.1, 12.2, and 12.3, the entropy
increases by $k\log 2$ during measurement; and in the special case illustrated
in Figures 12.4 and 12.5, entropy is conserved since the measure of all the
macrostates involved is the same. Can entropy also decrease during
measurement?

Suppose that O replaces the measuring device D by a better device D'
that can distinguish between the subregions of $[M_3]$ and $[M_4]$ that come
from $[M_1]$ and $[M_2]$. That is, the macrostates of D' that are correlated with
the different subregions of $[M_3]$ and $[M_4]$ are distinguishable by O. Such a
measuring device can help O know the original macrostate of G. In that
case, the above evolution will no longer constitute an erasure; and in
order to erase the information concerning the initial macrostates ($[M_1]$ or
$[M_2]$), we will need a different evolution, such as that illustrated in
Figure 12.6: each macrostate of D' contains both trajectories that came
from $[M_1]$ and trajectories that came from $[M_2]$.

The evolution illustrated in Figure 12.6 shows that the blending of
trajectories is relative to the *end macrostates* in the erasure evolution.
Since macrostates are relative to an observer, it follows that *erasure is
relative to an observer*. There is no universal erasure. There is no dynamics
that will bring about blending for all possible observers, namely for all
possible partitions of the phase space to macrostates, because such
dynamics would require that any region of positive Lebesgue measure
contain end points that arrive from any other such region after a finite

---

[5] Nor is it a theorem in quantum mechanics; see Appendix B.3.

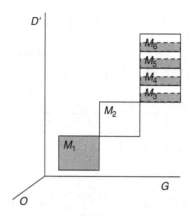

Figure 12.6 Entropy-reducing erasure

evolution time. However, for any given partition, there is a blending dynamics that brings about an erasure.

In the erasure set up in Figure 12.6 (as well as in all the previous set ups) Liouville's theorem is satisfied because the dynamics maps the union of the regions $[M_1]$ and $[M_2]$ to the union of the regions $[M_3]$, $[M_4]$, $[M_5]$, and $[M_6]$, which has the same total Lebesgue measure. However, the Lebesgue measure of *each* of the possible final macrostates is smaller than the Lebesgue measure of *each* of the initial macrostates. Thus, in this case, entropy decreases (provided that a measurement is carried out and a collapse takes place on the actual outcome. For simplicity, we shall not always add this clause henceforth).

And so, if the thermodynamic macrostates have a structure similar to the one in Figure 12.6, then this erasure is entropy-decreasing. Any claim to the effect that erasure is *necessarily* entropy-increasing (such as Landauer's thesis, on which we expand below) has to show that the structure of the thermodynamic macrostates is such that each of the macrostates $[M_3]$, $[M_4]$, $[M_5]$, and $[M_6]$ must be larger than each of the macrostates $[M_1]$ and $[M_2]$. We are not aware of any such demonstration.

12.6 Another logically irreversible operation

This result, that erasure does not have to involve entropy increase by any minimum amount and can even be entropy-decreasing, is not specific to the case of erasure but applies to the physical implementation of any

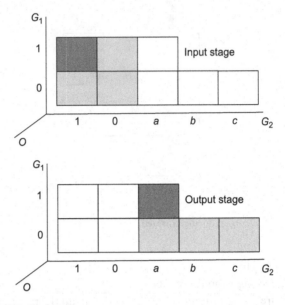

Figure 12.7 Conjunction

logically irreversible operation. Indeed, Landauer[6] was the first to point out, in 1961, the alleged connection between erasure and entropy in the context of the entropy cost of the physical implementation of logically irreversible operations in general. Erasure is the simplest of these operations, and it is perhaps for this reason that Landauer and others focused on it. The logical operation of conjunction is another example of a logically irreversible operation.

Conjunction is a function which has the following form: $C(1,1) = 1$, and $C(1,0) = C(0,1) = C(0,0) = 0$. It is logically irreversible since the output 0 is compatible with the three different inputs $(1,0)$, $(0,1)$, and $(0,0)$, and therefore an observer who has only the information about the output cannot know which of them was the actual input.

We will now show how conjunction can be physically implemented with no change of entropy. In Figure 12.7, the axes G_1 and G_2 correspond to two information-bearing degrees of freedom at the input stage, where G_1 stores the output. The input macrostate $(1,1)$ evolves according to the dynamics, so that by the end of the operation the dynamical blob that started out in $(1,1)$ is in region a (which has the same measure as regions

---

[6] Landauer (1961).

1 and 0) along G_2, and in the region 1 along the G_1 degree of freedom. The other input macrostates evolve according to the dynamics in such a way that by the end of the operation the dynamical blob that started in (1,0), (0,1), and (0,0) ends up in regions a, b, and c respectively (all of which have the same measure as 1 and 0) along G_2, and in the region 0 along the G_1 degree of freedom. (At the output stage one can of course use another degree of freedom instead of G_2.)

Since by construction all the macrostates in the figure have the same Lebesgue measure, Liouville's theorem is satisfied, and entropy (here understood as the Lebesgue measure of the macrostates) is conserved. Every instance of a physical implementation of the function C is carried out by a single trajectory that starts out somewhere in one of the initial macrostates (1,1), (1,0), (0,1), and (0,0), and ends at a point somewhere in one of the four macrostates $(a,1)$, $(a,0)$, $(b,0)$, and $(c,0)$, and all these instances are *entropy-conserving*.

The operation described above is a physical implementation of a logic-ally irreversible operation but it is of course physically reversible, even at the macroscopic level, since the information concerning the input is stored in the G_2 degree of freedom. However, it is possible to make this oper-ation macroscopically irreversible, and without losing the conservation of entropy, by using the blending method. For example, one could map 1/3 of each of the input macrostates (1,0), (0,1), and (0,0) to each of the output macrostates $(a,0)$, $(b,0)$, and $(c,0)$. In this way the operation will be logically irreversible, physically irreversible (at the macroscopic level), and yet entropy-conserving.

12.7 Logic and entropy: a model

As we saw at the beginning of this chapter, the conventional wisdom has it that the physical implementation of any logically irreversible operation brings about entropy increase in the environment by the minimum amount of $k\log 2$ per lost bit of information. This belief may have motivated the search for embedding the logically irreversible operations in logically reversible algorithms, such as Fredkin's gate.[7] However, since – as we have just showed – Landauer's thesis is false, these results, although interesting as theorems in logic, do not bear on the

---

[7] Fredkin and Toffoli (1982).

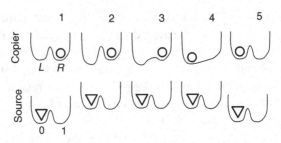

Figure 12.8 Copying: a bi-stable well implementation: case A

thermodynamics of computation. We now wish to consider a physical model which is taken in the literature as paradigmatic and as demonstrating Landauer's thesis, and show how exactly the supposed link between erasure and entropy increase is made.[8]

Let us start with a physical implementation of the logically reversible two-bit operation of copying, which is the function that maps (1,1) to (1,1) and (0,1) to (0,0), in which the copier starts in the Ready state 1 and ends with the same value as the source. In the model of the physical implementation of this function, both the source and the copier are bi-stable potential wells.[9] Consider Figure 12.8.

At Stage 1 in the figure the source is in state 0 and the copier in the ready state R. At Stage 2 the two are brought together, and the interaction between them is such that the potential well of the copier becomes deeper at the side of the source. At Stages 3 and 4 the potential barrier in the copier is gradually lowered by changing some of the external constraints on the copier and, because of the lowered L side, the potential well becomes tilted, and so the copier moves to the L side. The whole process is carried out slowly, so as to minimize friction, although some friction is necessary in order to prevent the copier from bouncing back to its original position at Stage 4. At Stage 5 the barrier is raised again and the source is removed from the copier.

If the source happens to start out in 1, the operation is as in Figure 12.9, where the external agent acts on the source and the copier in exactly the same way as before. The potential well grows deeper at R, tilts accordingly, and ends in its initial state.

Let us now examine the state-space representation of this process. We begin by assuming that the observer knows whether the source starts in

[8] This model is discussed by, for example, Bennett (1982) and Feynman (1996).
[9] Following Bennett (1982) and Feynman (1996).

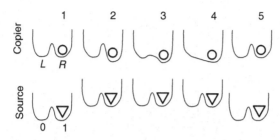

Figure 12.9 Copying: a bi-stable well implementation: case B

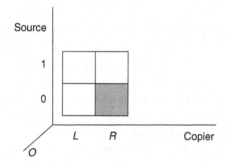

Figure 12.10 State space of the copying: case A

0 or in 1, and we will then proceed to the case where the observer does not have this information.[10] The latter case is important for the analysis of computers, in which often the states of memory cells are not measured during the computation process.[11]

Start with the case described in Figure 12.8. A statistical mechanical description of a system consists of three elements: the system's accessible region, the possible macrostates in which the system can be found by a given observer upon measurement; and the dynamics. Here, the accessible region of the source and the copier consists of the regions 0 and 1 of the source and of the regions L and R of the copier, as illustrated in Figure 12.10. All the microstates in these regions are compatible with the basic structure of the memory cells of the source and the copier.

The accessible region is partitioned into macrostates, as in Figure 12.10. The dynamics leads the system at each stage to the desirable state space region and, as we will see, it is controlled by changes in the structure of

[10] We follow Bennett (1982) in the distinction between these two cases.
[11] This is pointed out by Landauer (1961).

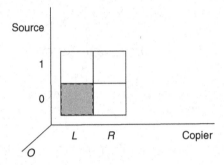

Figure 12.11 Evolution of the copier: case A

the potential well. We now turn to describe the details of the dynamics of the copying operation, in the state space. At the initial Stage 1, the source and the copier are in the initial macrostate [R0], as indicated in Figure 12.10, and the dynamics confines the system to remain in this region. This confinement is important since it ensures that the source and the copier store information in a stable way, and thus can be used for physical implementation of logical operations.

The interaction between the source and the copier at Stage 2 changes the details of the trajectories in the accessible region, but their overall structure remains unaltered, and the source and copier remain in macrostate [R0]. At Stage 3 the potential barrier is lowered by a suitable change of the external constraints, until at Stage 4 the potential well becomes tilted, at which point the structure of the trajectories undergoes a radical change: the dynamical blob of the source and copier evolves into macrostate [L0]. This evolution of the blob is brought about by the force field expressed in the tilted potential well. (As we said, at Stage 4 some friction is also necessary, in order to prevent the copier from bouncing back to R, but this friction can be minimized by keeping everything gradual and slow.[12])

At Stage 5 the potential barrier is raised, and this change of the external constraints on the copier brings about another change in the structure of the trajectories, which now confine the source and copier to macrostate [L0].

The case in which the system starts out in R is analogous: the initial macrostate is [R1], but in this case all the stages of the experiment are described by the same figure, namely Figure 12.12. The source and copier end up confined to macrosate [R1].

---

[12] This point is emphasized by Feynman (1996).

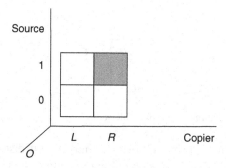

Figure 12.12 State space of copying: case B

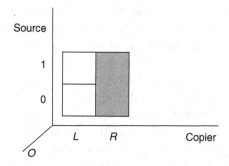

Figure 12.13 Copying random data

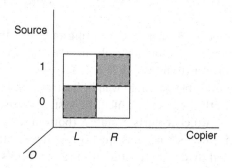

Figure 12.14 Evolution of copier of random data

Up to now, the description has portrayed the case in which before the copying takes place the observer *knows* the initial macrostate of the source, possibly by way of measuring it. As we said, in order to understand the physics of computers it is also important to consider the case where before the copying takes place the observer does not know which is the initial macrostate of the source. In that case, since a macrostate is

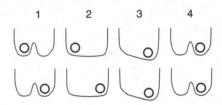

Figure 12.15 Erasure using a bi-stable well

defined as including all the microstates that are not distinguished by an observer, we would say that the initial macrostate is the union of [R0] and [R1], which we call [R0 + R1], as in Figure 12.13. (But the two parts of this macrostate are not interaccessible: if the actual microstate starts out in one of them it will not evolve into the other.) Following the lowering of the potential barrier and the tilting of the potential well, the dynamical blob that started out in [R0 + R1] evolves into the two regions shaded in Figure 12.14. If the observer measures the macrostate of the source and copier, it will (probably) be found in *either* [L0], *or* [R1]. This completes the operation of copying the state of the source to the copier.

According to the standard wisdom, copying does not necessarily involve any minimum of entropy increase. This is compatible with our analysis above, but it is crucial to note that in our analysis the behavior of entropy depends *contingently* on the measure of the macrostates, while there is no necessary connection between the measure of the macrostates and the physics of copying.

We now wish to describe, in a similar setting and using the same kind of reasoning, the simplest one-bit logically irreversible operation, namely erasure, which is the function $f(0) = f(1) = 1$. The physical implementation of erasure using a bi-stable potential well is illustrated in Figure 12.15[13] in which we depict both the case in which the system starts in state L (top row) and the case in which it starts in state R (bottom row). At Stage 2 the potential barrier is lowered and subsequently, at Stage 3, the R side of the potential well is slightly deepened, so that the system ends up somewhere in R, regardless of its initial state.[14] (Again, as in Stage 4 of the copying operation, at Stage 3 of the erasure some friction is necessary in order to prevent the system from bouncing back to L, but this friction can be

---

[13] Following Bennett (1982).
[14] It is sometimes said that entropy increases following the lowering of the barrier by analogy with the removal of a partition of a gas in a box. But the removal of the partition does not result in an instantaneous entropy increase since the macrostate of the gas is unchanged.

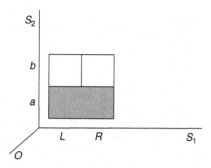

Figure 12.16 State-space description of erasure of random data: part 1

minimized by keeping everything gradual and slow.) Finally, at Stage 4, the potential barrier is raised again and the potential well is symmetrized back to its initial state.

The popular opinion has it that somewhere in Stage 2 or Stage 3 of this model an entropy increase of at least $k\log2$ *necessarily* takes place in the environment. Richard Feynman, in his *Lectures on Computation*, expresses this idea when he says that "The phase space of the inputs has shrunk to that of the output, with an unavoidable decrease in entropy. This must be compensated by heat generation somewhere."[15] Here Feynman repeats Landauer's thesis,[16] which is mistaken, as we have already shown. We will now show how our argument concerning this thesis works in the context of this particular model, pointing out where the notion of blending comes into play and what exactly determines the entropy changes during this process.

In Figure 12.16, S_1 is the information-bearing degree of freedom of the system and S_2 is another degree of freedom, either of the system or of its environment. Suppose, first, that initially the observer does not know whether the system is in L or in R of its bi-stable potential well (this is the case that Bennett[17] called "random (or unknown) data"). In this case, according to the definition of a macrostate, the initial macrostate is $[Ra + La]$, as in Figure 12.16, and the entropy is determined by the ν-measure of $[Ra + La]$. Notice that at Stage 1 (in Figure 12.15) the

---

[15] Feynman (1996, p. 153–4)
[16] Landauer (1961); here is what Bennett (1982), p. 934 says: "The irreversible entropy increase occurs at the point indicated by the arrow in Figure 16C, where the system's probability density leaks from the minimum it was originally in to fill both minima ... This is analogous to the free expansion of a gas into a previously evacuated container, in which the gas increases its own entropy without any work on its environment."
[17] Bennett (2003); from now on we refer in this section to this paper unless otherwise stated.

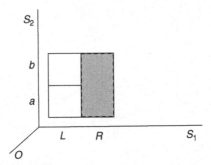

Figure 12.17 State-space description of erasure of random data: part 2

trajectories must be such that the two state-space regions *Ra* and *La*, which make up the macrostate [*Ra* + *La*], are *not interaccessible*: if the actual microstate starts in one of them it will not evolve into the other, in order for information to be stably stored in the system, so that we do indeed have a physical implementation of a logical operation.

At Stage 2, following the change of external constraints that lower the potential barrier, the *Ra* and *La* regions become dynamically interconnected, and the system can move between them. The tilting in Stage 3 brings about a dynamics in which the trajectories that start out in *Ra* will remain in *Ra*, and the trajectories that start in *La* will also *evolve* into the *R* region of S_1. In order for this evolution to satisfy Liouville's theorem, we must use the degree of freedom S_2, and map *Ra* and *La* onto the regions *Ra* and *Rb*, as in Figure 12.17. This consequence of Liouville's theorem is one of Landauer's important insights.[18]

This evolution has two important features. (i) The *entropy* change during this process is determined by the way in which the accessible region happens to be partitioned into macrostates, relative to the relevant measurements; and (ii) whether or not this constitutes an erasure depends on whether this evolution is blending, and this in turn depends on the partitioning into macrostates together with some details of the dynamics. There is a significant dispute in the literature about these features, and we now examine them in detail starting with (i).

In the first case that we examine here, the regions *a* and *b* of S_2 are undistinguishable from one another, so that the new macrostate of the system is [*Ra* + *Rb*], as depicted in Figure 12.17: since the observer does not (or cannot) become correlated with any finer regions along S_2, *O* is

[18] See Landauer (1961).

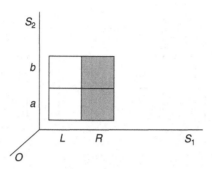

Figure 12.18 State-space description of erasure of random data: part 3

correlated with the entire region [*Ra* + *Rb*], which is the post-erasure macrostate. Moreover, this evolution is blending: from the new macrostate it is impossible to infer which half of the initial macrostate [*Ra* + *La*] contained the actual microstate. In this case entropy is conserved, since the measure of the final macrostate [*Ra* + *Rb*] is the same as the measure of the initial macrostate [*Ra* + *La*]. However, since the entropy along S_1 decreases, the entropy along S_2 increases by the same amount. The total entropy balance of the universe is zero. This is the case that Landauer had in mind; in particular he seems to have argued that the entropy increase along S_2 is *necessary*. However, Landauer overlooked the question of measurement which, as we argued, determines the final entropy.

Consider the second case, depicted in Figure 12.18. Here, the regions *a* and *b* of S_2 are distinguishable by the observer: this observer carries out a measurement of S_2 and collapses the blob onto one of the two macrostates [*Ra*] or [*Rb*]. In this case, the entropy of $S_1 + S_2$ *decreases*, relative to the pre-erasure macrostate [*Ra* + *La*]. It is crucial to note that the act of measurement, which in this case decreases the entropy of the universe, has no other entropic effects.

We now turn to feature (ii) above, namely the question of whether or not the dynamics in this case is blending (and hence erasing), which depends on the details of the evolution of the blob. And here there are two possibilities. One possibility is that the evolution is such that the dynamics maps *Ra* to itself and *La* to *Rb*, in which case by observing the macrostate of the system an observer will be able to retrodict which half of the initial macrostate [*Ra* + *La*] contained the actual microstate, and so this dynamics is not blending and not erasing. Another possibility is that the dynamics is blending: half of the trajectories that start in *Ra* are

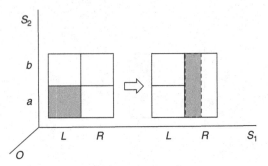

Figure 12.19 State-space description of erasure of known data: case A

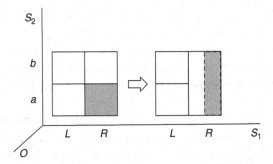

Figure 12.20 State-space description of erasure of known data: case B

mapped to *Ra* and half of them are mapped to *Rb*; and similarly for the trajectories that start in *La*: half of them are mapped to *Ra* and half of them are mapped to *Rb*. In this case, measuring the macrostate of the system – and in particular of S_2 – will not reveal whether the initial microstate of the system was in *L* or *R*.

Finally, we turn to consider a different version of the same erasure experiment in which the initial macrostate of the system is *known*, which means that it is known to be either [*Ra*] or [*Rb*] (see the left-hand parts of Figures 12.19 and 12.20). The time evolution resulting from the tilting is constrained by Liouville's theorem, and so the end points of trajectories that start in [*Ra*] cannot overlap with the end points of trajectories that start in [*Rb*]. And so, the dynamical blob that starts in any of the two initial macrostates in Figure 12.16 (described in the left-hand-side parts of Figure 12.19 and 12.20) must evolve into half of the blob appearing in Figures 12.17. However, to erase the known data one must have blending dynamics, such as that described

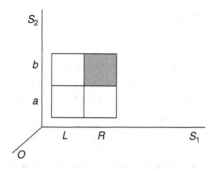

Figure 12.21 Post-erasure macrostate *A*

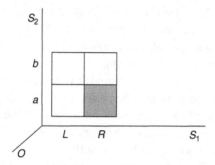

Figure 12.22 Post-erasure macrostate *B*

in the right-hand-side parts of Figures 12.19 and 12.20. If the observer now does not know the details of the mapping over and above the fact that the blob arrived into *Ra* + *Rb* (as in the right-hand side of the two figures), then the state of knowledge of this observer will be described by the entire region [*Ra* + *Rb*], which would be the macrostate in this case. And so in this scenario of erasure the total entropy of the universe would increase.[19]

However, this entropy increase is not necessary. Suppose, for example, that in the case of Figure 12.19 the observer measures on the post-erasure blob whether the system is in the macrostate [*Ra*] or [*Rb*]. The outcome after the collapse has taken place (that is, after the split, detection and expansion stages of the measurement) is either the one in Figure 12.21 or the one in Figure 12.22. Similarly for the case of Figure 12.20: here, too, the outcome will be either the one in Figure 12.21 or in Figure 12.22. And

---

[19] Note that this is different from the erasure of random data, and it is in line with, for example, Bennett (2003).

note that this measurement can be finished at the same time that the erasure (in Figures 12.19 and 12.20) ends. In these cases, *the post-erasure state has the same entropy as the pre-erasure state*. And of course other changes of entropy are possible, depending on the partition of the state space into macrostates at the same time (that is, the time at which the erasure is finished), according to the details of the correlations between the observer and the system.

12.8 What does erasure erase?

As we said in Chapter 10, the content of memory supervenes on some physical features of the *microstate* of the observer O. Since determinism implies that there is no microscopic erasure, the question arises: what precisely does the macroscopic erasure erase? It is clear that blending does not erase the memory of O. Blending only prevents *inferring* the actual macrostate in the past $[M_1]$ or $[M_2]$ from the present macrostate regardless of whether or not O remembers the past macrostate. Therefore blending prevents an inference from the present macrostate to the past macrostate only for an observer who has no memory of the past. In this sense the notion of memory erasure may be misleading.

Of course, as we know from experience, we do forget things about our past and (since this forgetting cannot be explained by blending because of essential determinism) forgetting means roughly the following. The set of all microstates of, say, the brain states of an observer are partitioned into two subsets: a subset of microstates that somehow generate memory *as a mental state* and the subset of microstates that do not have this property. In extreme (but not necessarily infrequent) cases, the microstate of the brain evolves into a microstate of the kind that generate no mental state, such as the microstate of a dead brain. Each of the microstates in the first subset is a mental state with some content. Forgetting means that a microstate in the first subset is mapped into the second. But again this mapping has nothing to do with blending, that is, with erasure of the kind that is considered by Landauer's thesis.

12.9 Conclusion

A careful analysis of the physical implementation of logical operations, in which the details of the accessible region, the macrostates, and the dynamical evolution are addressed, reveals that there is no necessary connection between logical properties of logical operations and the physical property of entropy of their physical implementation.

The implications of this counter-mainstream conclusion for the physics of computers are not dramatic, since entropy change by $k\log2$ per bit is exceedingly small relative to the dissipation that takes place in actual computers. At the same time, the theoretical significance of this conclusion should not be underestimated, in at least two contexts. First, this conclusion entails that whether or not logic and mathematics are constrained by physics is still an open question.[20] Since Landauer's thesis is not a theorem in mechanics, it seems to us that this linkage depends on the precise details of the physical implementation of the logical operations rather than the fundamental laws of physics. And second, this conclusion opens the way to proving that a Maxwellian Demon, which is a counter-example to the probabilistic counterpart of the Law of Approach to Equilibrium, is consistent with mechanics. We finally come to the end of our journey, in which we describe the Demon and show how it works.

[20] Landauer (1992, 1996).

13

Maxwell's Demon

13.1 Thermodynamic and statistical mechanical demons

In 1867 James Clerk Maxwell wrote to Peter Guthrie Tait that there was a "hole" in the Second Law of Thermodynamics. Maxwell wrote:

Now let A & B be two vessels divided by a diaphragm and let them contain elastic molecules in a state of agitation which strike each other and the sides.

Let the number of particles be equal in A and B but let those in A have greatest energy of motion. Then even if all the molecules have equal velocities, if oblique collisions occur between them, their velocities will become unequal, and I have shown that there will be velocities of all magnitudes in A and the same in B, only the sum of the squares of the velocities is greater in A than in B.

When a molecule is reflected from the fixed diaphragm CD no work is lost or gained.

If the molecule instead of being reflected were allowed to go through a hole in CD no work would be lost or gained, only its energy would be transferred from one vessel to the other.

Now conceive a finite being who knows the paths and velocities of all the molecules by simple inspection but who can do no work except open and close a hole in the diaphragm by means of a slide without mass.

Let him first observe the molecules in A and when he sees one coming the square of whose vel. [velocity] is less than the mean sq. [square] velocity of the molecules in B let him open the hole and let it go into B. Next let him watch for a molecule of B, the square of whose velocity is greater than the mean sq. [square] vel. [velocity] in A, and when it comes to the hole let him draw the slide and let it go into A, keeping the slide shut for all other molecules.

Then the number of molecules in A and B are the same as at first, but the energy in A is increased and that in B diminished, that is, the hot system has got hotter and the cold colder and yet no work has been done, only the intelligence of a very observant and neat fingered being has been employed.

Or, in short, if the heat is the motion of finite portions of matter and if we can apply tools to such portions of matter so as to deal with them separately,

270

then we can take advantage of the different motion of different proportions to restore a uniform hot system to unequal temperatures or to motions of large masses.

Only we can't, not being clever enough.[1]

Maxwell continues, in a subsequent letter:

Concerning Demons

1. Who gave them this name? Thomson.
2. What were they by nature? Very small BUT lively beings incapable of doing work but able to open and shut valves which move without friction or inertia.
3. What was their chief end? To show that the 2nd Law of Thermodynamics has only statistical certainty.
4. Is the production of an inequality of temperature their only occupation? No, for less intelligent demons can produce a difference in pressure as well as temperature merely by allowing all particles going in one direction while stopping all those going the other way. This reduces the demon to a valve like that of the hydraulic ram, suppose.[2]

Maxwell takes his Demon to be more "clever" than us in the sense of having better observation capabilities and finer manipulation capabilities; but it is, nevertheless, a physical device, no less than a "valve." Nowadays we would perhaps use the term robot rather than demon or valve. There is nothing supernatural or non-physical about it. We take the notion of a natural Demon to mean that the Demon consists of particles that obey the laws of physics. It may and should be represented in terms of the degrees of freedom of its particles just like an ideal gas in a container.

Maxwell thought of his Demon as showing that thermodynamics is incompatible with mechanics, since mechanics is compatible with evolutions that are entropy-reducing. He believed (so it seems, see point 3 in his second letter above) that the Law of Approach to Equilibrium is true provided that it is understood probabilistically, that is, as saying that systems are highly likely to satisfy this law. That is, anti-thermodynamic evolutions are possible, but improbable. In the years that followed Maxwell's letter to Tait, the theory of statistical mechanics has been developed, and it is now widely believed that Maxwell's intuition concerning the improbability of anti-thermodynamic evolutions is correct: it is widely held that whereas an *anti-thermodynamic Demon* is possible, an *anti-statistical mechanical Demon* is impossible.

---

[1] Knott (1911, pp. 213–214),
[2] Knott (1911, pp. 214–215).

This prevalent opinion has been supported by numerous arguments[3] which proposed a variety of models of the Demon; each emphasizes a different aspect of the thought experiment, which the writer thought would be the key to showing why the idea of a Demon is fundamentally mistaken and therefore a system that systematically violates the thermodynamic regularities is impossible. However, since particular models always contain many subtle technical details, and imperceptibly assume many other details, these models have led only to confusion.[4] A way out of this muddle has been offered by David Albert in his book *Time and Chance*,[5] where instead of offering yet another particular model, Albert turned to a general state-space analysis of the Demon. In his general analysis Albert challenged the (until then) universally held opinion (with the possible exception of Maxwell), and proved that a system that decreases the entropy of the universe with *high* probability is compatible with classical mechanics.[6]

Our discussion in this chapter follows and extends Albert's argument in the most general terms, and refrains from examining particular devices. We will argue that a Maxwellian Demon is compatible not only with the principles of mechanics, but also with the principles of statistical mechanics. Whether or not a Demon can be realized in experience is an open question, and the answer to it depends on the details of the dynamics and the initial condition of our world. As of now, as we emphasized throughout the book, no general proofs or observations that settle the question one way or the other are available, despite strong intuitions. In what follows we first embed Albert's argument in the general conceptual framework put forward in this book, namely, using the idea that probability in statistical mechanics emerges as a consequence of the interplay between macrostates and blobs. Here we use the notion of measurement as developed in Chapter 9. We then complete the Demon's cycle of operation, by erasing its memory and returning it to its initial state; here, we use the understanding of erasure proposed in Chapter 12. And finally, in Appendix A we

---

[3] See Leff and Rex (2003). See also Earman and Norton (1998, 1999).
[4] One example of a confusion that is worth mentioning, because of the popularity of the argument, concerns Leon Brillouin's influential black body argument (e.g., article 3.4 in Leff and Rex 2003). This argument is mistaken (apart from its shortcomings pointed out by Leff and Rex in pp. 17–19) since, in all the versions of his argument, Brillouin forgets to take into account the fact that when the energy packet enters the Demon's eye it also leaves the gas. Adding the missing negative quantity makes the total entropy balance negative (thus Brillouin seems to disprove his own conclusion).
[5] Albert (2000, Ch. 5).
[6] Maxwell's Demon is compatible also with quantum mechanics; see Appendix B.3.

illustrate these ideas in the special case of a particle in a box (known as Szilard's engine), and generalize the argument by showing that a Demon is also compatible with quantum mechanics in Appendix B.3.

13.2 Szilard's insight

In his 1929 paper, "On the Decrease in Entropy in a Thermodynamic System by the Intervention of Intelligent Beings,"[7] Leo Szilard suggested that the key to understanding Maxwell's Demon is what he called "intelligence": in his 1929 paper the focus was on *measurement*, namely on acquiring information. Later on, Bennett applied Landauer's idea, concerning the entropy cost of logically irreversible operations, to the study of the Demon, and the focus in the literature was transferred to loss of information by erasure. In Chapters 9 and 12 we have shown that a careful analysis of operations associated with "intelligence" in any of the above senses reveals that they have no particular significance as far as entropy is concerned. That is, it turns out that the standard ideas concerning the essential aspects of Maxwell's Demon are mistaken. Nevertheless, we take it that Szilard's insight focusing on the intelligence of the Demon remains valid. Moreover, in our view the best way to understand Albert's argument is by following Szilard's insight. That is, the Demon reduces the entropy of the universe in a suitable measurement and then completes the cycle of operations including erasing its own memory, without any further entropy increase anywhere in the universe. The entropy changes during these operations depend only on the particular details of the physical set up in which they are implemented.

An unusually short proof that a Maxwellian Demon along these lines is consistent with statistical mechanics is obtained by combining Figures 9.2, 9.3, and 9.6 describing an entropy-decreasing measurement together with Figures 12.19, 12.20, 12.21, and 12.22. The remainder of this chapter gives further details.

13.3 Entropy reduction: measurement

Let us now introduce a Maxwellian Demon. Figure 13.1 depicts the players in the Maxwellian Demon scenario: G is a thermodynamic system, such as an ideal gas in a container; D is a measuring device; and E is the environment, which includes the rest of the universe except for the

---

[7] See Szilard (1929).

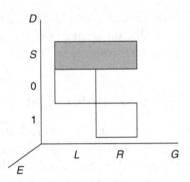

Figure 13.1 Initial macrostate of a Demonic universe

observer O, relative to which the macrostates of G, D, and E are determined. We can think of O as a human observer who cannot distinguish
between the L and R set of microstates of G by a direct observation of G;
but D can distinguish between them, and so O can use D as a measuring
device of G, and distinguish between L and R by observing the outcomes
of measurements that D carries out on G. The observer O does not appear
in the figure since we will need to depict only the three elements D, G, and
E; we will describe O's state as we proceed.

Initially, at time t_0, D is in its Ready macrostate S, and so O cannot
distinguish between L and R and concludes that the macrostate of $D +G$ is
$[S, L + R]$, as in Figure 13.1. The environment E is initially in a macrostate
$[e_1]$ which, for simplicity, we do not describe at this stage.

At t_1 O starts to carry out a measurement of the gas G by the measuring
device D. As usual, the first stage of the measurement is the dynamical
split: the dynamics takes the dynamical blob $B(t_1)$ of $D + G$ to regions that
overlap with macrostates $[0, L]$ and $[1, R]$, as in Figure 13.2. Only one of
the microstates in $[S, L + R]$ is actual, and hence only one of the microstates in $[0, L]$ and $[1, R]$ is actual.

This stage (and the entire process) satisfies Liouville's theorem: the
Lebesgue measure of the dynamical blob $B(t_1)$ at t_1 is the same as the
Lebesgue measure of the initial macrostate $[S, L + R]$ at t_0. For clarity of
presentation we chose each of the macrostates $[1]$ and $[0]$ of D to be a little
larger (in Lebesgue measure) than the initial macrostate S, and so
the union of $[0, L]$ and $[1, R]$ is a little larger than the initial macrostate
$[S, L + R]$, and therefore it is easier to see how the blob only *partly*
overlaps with the two macrostates. But other cases are possible, such as
that in which the blob exactly overlaps with the two macrostates, as in
Figure 13.3. Although this latter case is popular in the literature on the

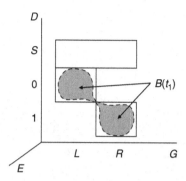

Figure 13.2 Demonic dynamical evolution

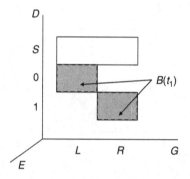

Figure 13.3 Case of exact overlap

Demon,[8] it is a special case, and we prefer to focus on a more general one, i.e. that in Figure 13.2. Our argument, phrased in terms of Figure 13.2, is fully applicable for the case in Figure 13.3 as well.[9]

At time t_2, the measurement of G by D ends with a collapse on the macrostate of $D + G$ that corresponds to their actual microstate, which is microscopically correlated with O's actual microstate at t_2. The macrostate of $D + G$ by the end of the measurement is either $[0, L]$, as in Figure 13.4, or $[1, R]$, as in Figure 13.5. While the transition probability to each

---

[8] For example Bennett (1982) and Fahn (1996)
[9] In the special case where the blob *exactly* overlaps with the post-measurement macrostates, the velocity reversal of each of the end microstates returns to the initial macrostate in the same time interval. This leads to a problem of efficiency which we discuss below. In the more general case, where the measure of the union of the post-measurement macrostates is larger than the measure of the initial macrostate, the efficiency of the Demon can be improved, but there is a trade-off between the amount of entropy decrease and predictability.

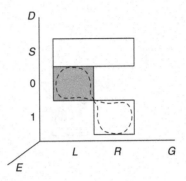

Figure 13.4 Collapse on outcome [0, L]

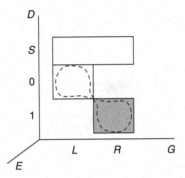

Figure 13.5 Collapse on outcome [1, R]

of these final macrostates given the initial macrostate depends on the μ-measure of the overlap between the blob $B(t_1)$ and the two macrostates, the probability that $D + G$ will end on one of them is strictly 1: no other macrostate is dynamically possible at t_1 and at t_2. Note that this construction is compatible with Loschmidt's reversal (see Section 7.8).

As we explained in Chapter 9, this measurement is entropy-decreasing in the technical sense, assuming that the ν-measure of the final macrostate (whether it is [0, L] or [1, R]) is smaller than the ν-measure of [S, L + R]. And since the probability of ending in either [0, L] or [1, R] is 1, the probability of entropy decrease is 1. That is, in this arrangement all the microstates in the initial macrostate lead with certainty to a decrease of the total entropy including microstates that are velocity reversals of each other. This is why this construction is immune to Loschmidt's reversibility objection.

This decrease does not violate Liouville's theorem, since trajectories are not cut, nor do they branch in. The dynamics of the measurement is such that the observer O simply becomes correlated with the microstates in the final macrostate at the end of the measurement. In this sense one can say metaphorically that when the measurement is completed O realizes that the counterfactual macrostate is irrelevant for further predictions and ceases to take it into consideration. This is the essence of the idea of *collapse* in classical measurement.

We draw attention to the connection between our state-space construction of a Demonic universe, and Maxwell's thought experiment. Notice that in our construction the entropy of G decreases with certainty, while the entropy of D is approximately constant (as we said, in Figure 13.2 entropy increases slightly, for clarity of presentation, but this increase has *no minimum* and can be made smaller and smaller until it vanishes as in Figure 13.3). This corresponds to Maxwell's original idea, in which demons are "very small BUT lively beings incapable of doing work but able to open and shut valves which move without friction or inertia."[10]

In our construction, the counterpart of Maxwell's Demon is essentially the observer, O. The role of O in this set up is solely to fix the partition of the state space into the thermodynamic macrostates. In this way we make explicit the distinction between the time evolution of the blob $B(t)$ and the description of the evolution of $D + G + E$ in terms of macrostates, a distinction which we know by now is crucial to the understanding of statistical mechanics. When we say that $G + D + E$ is initially in the macrostate $[S, L + R, e_1]$, we mean that its microstate is correlated with a microstate of O in a way which gives rise to this macrostate. This is what we mean (see Chapter 9) by a preparation of a system in a macrostate. We stress that the entire evolution of the universe *including* O is completely dictated by the dynamical structure of the trajectories.

Maxwell's picturesque Demon is sometimes treated as if it has *free will*, since it appears to place itself on the right trajectory in the state space that leads to a desirable state. In this sense the Demon appears to be a case of bootstrapping.[11] In our construction, there is neither bootstrapping nor any sort of goal directedness, since everything is a result of the details of the dynamics *including* the preparation (in the above sense) of the initial macrostate. Finally, in order to display a Demonic behavior as in the above scenarios, the universe must start out in the particular macrostate $[S, L + R]$. Once it is in this macrostate the universe will evolve Demonically, i.e. spontaneously in the way in which we described above. But how

---

[10] See Knott (1911, pp. 214–215).
[11] See Shenker (1999).

can the universe arrive at this initial macrostate, in the first place? It is a consequence of Liouville's theorem that the Lebesgue measure of [S, L + R] cannot be greater than half of the measure of the entire accessible region, and therefore this macrostate is not the one often taken to correspond to thermodynamic equilibrium as, for example, in Boltzmann's construction where the volume of an equilibrium macrostate (especially the Maxwell–Boltzmann macrostate) takes up almost the entire accessible region in the phase space (see Chapters 5 and 7). However, since our actual universe is in a low-entropy state right now, this constraint does not really undermine the possibility that our universe will evolve Demonically in the future.

13.4 Efficiency and predictability

In the above scenario of the Demon there is a trade-off between a reliable entropy decrease and macroscopic predictability. We want now to draw another connection, namely one between the predictability of the Demonic evolution and the efficiency of the Demon in reducing entropy. By the efficiency of an operation we mean the entropy difference between the initial and the final macrostates.

Consider one of the microstate points in the macrostate [S, L + R]; call it x (see Figure 13.6). In the evolution, as described above, this point must sit on a trajectory that takes it to a microstate (call it y) in one of the macrostates [0, L] or [1, R], say after τ seconds. Consider now the microstates which are the velocity reversals of x and y, call them $-x$ and $-y$ respectively. In many interesting cases (but certainly not in all cases) microstates that are the velocity reversals of each other belong to the

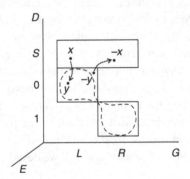

Figure 13.6 Loschmidt reversal

same macrostate (see Chapter 7). Suppose that both x and $-x$ belong to $[S, L + R]$, and similarly, that both y and $-y$ belong to $[0, L]$. This puts constraints on the efficiency and macroscopic predictability of the Demonic evolution. Let us see why.

Since the dynamics is invariant under velocity reversal, if the trajectory starting in x in $[S, L + R]$ takes the universe (consisting of D, G, E, and O) to the microstate y in $[0, L]$, then the trajectory that starts in $-y$ in $[0, L]$ takes S back to the microstate $-x$ in $[S, L + R]$ after τ seconds. As the mapping from x in $[S, L + R]$ to y in $[0, L]$ reduces the entropy of the universe, the reversed evolution from $-y$ in $[0, L]$ to $-x$ in $[S, L + R]$ increases the entropy of the universe. However, if the universe starts in $[S, L + R]$ and evolves into, say, $[0, L]$ (thereby decreasing its entropy), we want it to remain in the low-entropy state $[0, L]$, avoiding points such as $-y$ which take the universe back to the higher-entropy state $[S, L + R]$ after τ seconds. If we wish to make the universe remain in $[0, L]$, we can take one of the following actions:

We can increase the volume of each of the post-measurement macrostates (while keeping their number fixed) and thereby increase their *total volume*. In this case, the relative measure of the set of $-y$ points in each of the final macrostates will decrease, and so the probability of returning to the initial macrostate will similarly decrease. The reason is that each of the final macrostates will include longer trajectory segments which map each of them to itself. But the larger the volume of each of the final macrostates, the smaller is the entropy difference between the initial and final macrostates. Here, there is a trade-off between the efficiency of the Demon (i.e. the amount of entropy decrease) and the *stability* of the low-entropy state.

We can increase the *number* of the final macrostates (given a fixed measure of each of them), so that each macrostate will still have a small volume (relative to the volume of the initial macrostate), but the total volume of the final macrostates will increase. In this case, the measure of the trajectories that arrive at each of the final macrostates will be small relative to its volume. The entropy of the universe will decrease in every cycle of the operation; moreover, the low-entropy final macrostate will be relatively stable. That is, we can divide the macrostates $[0, L]$ and $[1, R]$ into smaller (by the relevant measure) macrostates, and this would effectively mean that the entropy decrease is larger than in the original measurement. However, as the number of the final macrostates increases, the macroscopic evolution becomes less predictable. So there is a trade-off between the stability of the low-entropy state and the macroscopic *predictability*.

According to the optimal interplay between the three factors, stability, predictability, and efficiency, we can combine the above strategies, and so increase the total measure of the final macrostates *and* their number.

13.5 Completing the cycle of operation: erasure

The operation of the Demon as described by Albert ends at t_2, with the reduced entropy, claiming that this already constitutes a counter-example for the probabilistic counterpart of the Law of Approach to Equilibrium and the Second Law of Thermodynamics.[12] Other writers claim that in order to challenge these laws one has to carry on with the evolution until a cycle of operation is *closed*.[13] Without going into the question of whether a process that ends at t_2 is indeed sufficient to challenge these laws (defined in Chapter 7), we now proceed to show how the cycle of operation can be completed.

What is a completion of an operation cycle? A full closing of an operation cycle would mean that the universe returns to its initial state. But here no such periodic evolution is intended since we specifically want the entropy of the gas to remain lower than it was initially. Instead, the idea is that at the end of the cycle the situation will be as follows: G will end with entropy lower than its initial entropy. D will end in its original initial macrostate and, *ipso facto*, retain its initial entropy. The environment E must retain its initial entropy but it may end in a macrostate that is different from its initial macrostate.[14] Moreover, the overall final macrostate of $D + G + E$ must be such that a subsequent operation cycle that reduces the entropy can start, and once the second operation is completed, another one can start, and then another, perpetually. These conditions are often stated in terms of the following properties that the final macrostate of $D + G + E$ should have:

---

[12] Albert (2000, Ch. 5)
[13] For various formulations of the Second Law, see Uffink (2001).
[14] This latter requirement is in accordance with the standard literature: for example, writers who think that a Demon is impossible claim that completing the cycle of operation involves an increase of entropy in the environment, and therefore the environment's final macrostate is *a fortiori* different from its initial macrostate. While for these writers not only the macrostate of the environment changes, but the entropy of the environment increases, we allow for a different macrostate, but with the same entropy (see, for example, Szilard 1929, Bennett 1982).

(i) D must be in its initial Ready macrostate $[S]$.

(ii) G's macrostate must have the same v-measure of entropy as $[L]$ or $[R]$ (depending on the outcome of the measurement).

(iii) E's macrostate must have the same v-measure of entropy as $[e_1]$.

Conditions (i), (ii), and (iii) together entail that the v-measure of entropy of the final macrostate must be smaller than the v-measure of entropy of the initial macrostate. In addition, the following condition (iv) is often made:

(iv) The final macrostate of $D + G + E$ must be such that any memory of the measurement's outcome at t_2 must be erased, in the sense that the final macrostate must contain no macroscopic records of whatever sort that will allow retrodicting which state among the $[0, L]$ and $[1, R]$ was the actual macrostate of $D + G + E$ before the erasure. As we explained in Chapter 12, this means that the erasure dynamics must be *blending*. The requirement of erasure refers to the macroscopic level, since in classical mechanics there is no microscopic erasure.

Let us see how all the conditions (i)–(iv) can be satisfied. Here we will need to take into account the environment E in addition to D and G, and therefore we illustrate the process in three dimensions. Whether the measurement ends in $[0, L]$ or $[1, R]$, the trajectories that start in both macrostates need to be taken into account, in order to ensure that trajectories will not branch in. Since D needs to end in $[S]$ and since G needs to end in some macrostate $[A]$ that has the same v-measure as that of $[L]$ and $[R]$, the first attempt would be to map the union of $[0, L]$ and $[1, R]$ to the macrostate $[S, A, e_1]$, as in Figure 13.7. But this evolution violates Liouville's theorem.

The only way to keep G in $[A]$ and D in $[S]$ is to increase the Lebesgue measure of E's macrostate, and so the dynamical blob $B(t_3)$ at t_3 must overlap with (or be contained in) the macrostate $[S, A, e_2]$, illustrated in Figure 13.8. The Lebesgue measure of the post-erasure macrostate $[S, A, e_2]$ must be at least as large as the Lebesgue measure of the union of $[0, L]$ and $[1, R]$. For simplicity let us assume that they are exactly equal. (This is a case analogous to the special case which is prevalent in the literature; it is illustrated in Figure 13.3. A more general case would be one in which the Lebesgue measure of $[S, A, e_2]$ is *slightly* larger, but there is no minimum to this increase.) Usually,though not necessarily, this means that the v-measure of entropy of E also increases. Let us suppose that this is the case, which means that we have not yet satisfied condition (iii).

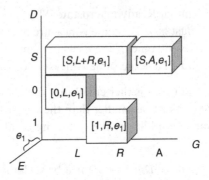

Figure 13.7 Erasure violating Liouville's theorem

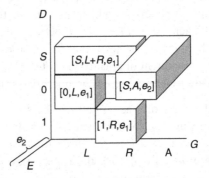

Figure 13.8 Entropy-increasing erasure

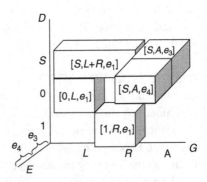

Figure 13.9 Entropy-conserving erasure

To satisfy condition (iii) we divide $[e_2]$ into two parts, call them $[e_3]$ and $[e_4]$, as illustrated in Figure 13.9, so that the v-measure of each of $[S, A, e_3]$ and $[S, A, e_4]$ is equal to the v-measure of $[0, L, e_1]$ and $[1, R, e_1]$. We also assume that these two parts are macrostates of E, distinguishable by the

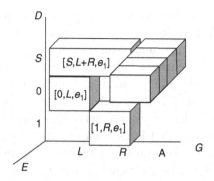

Figure 13.10 Entropy-decreasing erasure

observer O (we can think of no mechanical reasons to reject this assumption). Here we must be careful: if, for example, $[0, L, e_1]$ is mapped to $[S, A, e_3]$ and $[1, R, e_1]$ is mapped to $[S, A, e_4]$, this mapping will not be blending and will not constitute an erasure, and so requirement (iv) will not be satisfied. To satisfy (iv) we must have a blending dynamics, for example: half of the trajectories that start in $[0, L, e_1]$ evolve into $[S, A, e_3]$ and half into $[S, A, e_4]$, and similarly for the trajectories that start in $[1, R, e_1]$.

A blending dynamics that ends in the union of $[S, A, e_3]$ and $[S, A, e_4]$, as in Figure 13.9, satisfies all the four requirements (i)–(iv), and completes the cycle of operation. The final entropy is either the v-measure of $[S, A, e_3]$ or the v-measure of $[S, A, e_4]$, both of which are (presumably) smaller than the v-measure of the initial macrostate $[S, L+R, e_1]$. We could, by the way, end with even lower entropy, if the partition of macrostates had been as shown in Figure 13.10.

One point remains to be clarified again concerning the interplay between entropy decrease and predictability. By the end of each cycle there are two possible macrostates of E in the set up shown in Figure 13.9 (or four possible macrostates of E in Figure 13.10). Until O actually observes the final macrostate, there is no way of knowing which of the two possible final macrostates actually obtains. Whereas the final macrostate in the set up in Figure 13.8 is such that the entropy decrease during the measurement is lost at the erasure, the final macrostate is known with certainty to be $[S, A, e_2]$; the converse is the case in the set up shown in 'Figures 13.9 and 13.10. Here entropy is not lost, but macroscopic ignorance enters, and can be removed only by another measurement of E by O. But since measurements do not necessarily involve any entropic cost (as we proved in Chapter 9), this is absolutely irrelevant to the possibility of Maxwell's Demon. Generally, our construction shows that an exponential increase in the number of possible macrostates of E by the end of each cycle is perfectly compatible with a reliable and regular and repeatable entropy decrease and genuine memory erasure.

In sum, the entropy changes during the Demonic evolution are determined by the partitioning of the state space into macrostates, which are determined by the correlations between O and the rest of the universe: D, G, and E. Nothing in mechanics prevents the partitioning we used in this model, or the dynamics that leads from one stage to the next. By this construction we have demonstrated that the cycle of operation in a Demonic evolution can be completed, in the right sense of completion, as stated above. We have a Maxwellian Demon that is *consistent* with classical statistical mechanics.

13.6 The Liberal Stance

The Demon above rests on the idea that the partition of the state space into the thermodynamic macrostates is accepted as given, while we constructed a special dynamics that does the trick. However, we can think about the question of the Demon differently. We can assume the usual dynamics and change the partition into macrostates instead.

Suppose that, as a matter of fact, the world generally satisfies the Law of Approach to Equilibrium and the Second Law of Thermodynamics. That is: suppose that the dynamics of the universe is such that systems around us tend, with high *probability* (relative to the measure μ), to evolve into macrostates with increasing *entropy* (relative to the measure v) and once they arrive at the macrostate with the highest entropy they are highly likely to remain there. Although, as we have emphasized throughout this book, there is as yet no general proof that this is indeed the case, let us assume that our inductive generalizations hold, and that systems will continue to exhibit the observed regularities described by thermodynamics.

Given this reasonable assumption, it is still conceivable, and consistent with the laws of mechanics, that there exists a non-human observer, a *Demon* if you like, that is correlated with its environment in a different way from us, and consequently this Demon experiences the world in terms that are different from our macrostates. Relative to the set of macrostates experienced by this Demon, the dynamics of the universe may not satisfy the thermodynamic regularities that we experience. Nothing in mechanics precludes this possibility, and in this sense it is a possibility worth considering. We call this idea that there may be other observers, who experience the world in terms of macrostates that are different from ours, the *Liberal Stance* towards macrostates. Such a Demon can manipulate the world in a way that would appear to us as non-thermodynamic.[15]

A special case of such a Demon is simply a system that has observation capabilities that are finer than ours, namely, a system that can distinguish between sets of microstates that are imperceptible to our senses. We could use this Demon as a measuring device to give us information that would otherwise be inaccessible to us. And we could even use this Demon to manipulate systems according to these fine measurement outcomes, in a way that would be anti-thermodynamic. Of course, *if* the thermodynamic regularities were universal and inviolable for any partition to macrostates, *then* that would be impossible.[16] However, since there is no general proof from mechanics that such a scenario is impossible, and indeed we see nothing in mechanics that could prove it to be impossible, we think that this scenario is possible according to the principles of mechanics and therefore it is a statistical mechanical Maxwellian Demon. Using this scenario one may interpret the construction above in terms of Demonic macrostates and normal dynamics. Therefore, Maxwellian Demons are a consequence of the right sort of harmony between dynamics on the one hand and the partition into macrostates on the other. Note that whether or not the laws of thermodynamics hold is also a consequence of a harmony between the partition into thermodynamic macrostates and the dynamics.

The idea that Maxwell's Demon is consistent with classical mechanics has implications for the understanding of the meaning of the notion of entropy and the v-measure of entropy. As we saw in Chapter 7, the v-measure is chosen precisely in order to express the thermodynamic regularities. But if these regularities can be systematically violated by Demonic evolutions, there would be no connection between entropy and exploitability of energy. This of course does not undermine our proof of Maxwell's Demon.

---

[15] Grunbaum argues that "for any specified ensemble there will plainly be coarse-grainings that make the ensemble's entropy do whatever one likes, at least for finite time intervals." See Sklar (1993 p. 357).
[16] Note that this is essentially Albert's dynamical hypothesis (see Appendix B.1).

13.7 Conclusion

The law that entropy always increases, – the second law of thermodynamics – holds, I think, the supreme position among the laws of Nature .... [I]f your theory is found to be against the second law of thermodynamics I can give you no hope; there is nothing for it but to collapse in deepest humiliation.[17]

We believe that the attitude expressed here by Eddington is wrong. We have just shown that Maxwell's Demon is compatible with classical statistical mechanics, and we believe that there are some very good reasons to think that the general features of statistical mechanics make up a very good theory of the world; and therefore if statistical mechanics disagrees with thermodynamics, then one should not hastily give up statistical mechanics. We add here – without going into detailed arguments which are beyond the scope of the present book – that the transition from classical mechanics to quantum mechanics makes (very) little difference in this respect. The Demon also prevails in quantum mechanics. We give an example of a construction of a quantum mechanical Demon in Appendix B.3.

Nevertheless, if we take our experience as a guide, we cannot construct Demons. How can we explain this fact – given that Maxwellian Demons are, in principle, possible? As we said in the introduction, whether or not Maxwellian Demons can be constructed in our world depends on the kinds of Hamiltonians we can construct and the initial conditions we can control. Maxwellian Demons are possible in cases where there is the right sort of harmony between the dynamics (the evolution of the dynamical blobs) and the partition of the state space into macrostates. One can construct Demons either by finding the right sort of dynamical evolution to match a given set of macrostates (by constructing the Hamiltonians), or by finding the right set of macrostates to match a given dynamics (by constructing the right measuring devices). If we could achieve such a Demonic harmony, we could extract work from heat.

It seems to us that the difficulties in actually constructing a Demonic system involve practical issues such as controlling a large number of degrees of freedom of chaotic systems and their initial conditions, and isolating systems from external effects.[18] Since the issues involved here are

---

[17] Eddington (1935, p. 81).
[18] In quantum mechanics, interventionist constraints would presumably be related to decoherence effects. On the role of decoherence in statistical mechanics, see Hemmo and Shenker (2001, 2003, 2005).

merely practical, ruling out the possibility of a Demon in advance on the basis of the experience expressed in the Law of Approach to Equilibrium and the Second Law of Thermodynamics is circular reasoning. And the fact that Demons have not been observed in the past certainly does not rule out the possibility that Demons will not be observed or constructed in the future.

It is hard to settle the question of whether the thermodynamic regularities have not been violated until now (as far as we know) just because they are strictly true as a matter of physical law, or because their violation requires control that is much beyond our capabilities. And so the question of whether Demons are possible in our world, which is the question of whether the Law of Approach to Equilibrium and the Second Law of Thermodynamics are true in our world, remains open. To settle this question it seems that we must wait for the final theory of everything. Until then, we are willing to help Lawrence Sklar in helping Joel Lebowitz in being the bankers who "keep the money until everyone has agreed on the matter."[19]

[19] Sklar (1993, p. 420).

Appendix A
Szilard's engine

In Chapter 13 we formulated our general state-space argument according to which Maxwell's Demon is consistent with classical statistical mechanics. This is the general argument in support of Maxwell's intuition. We now wish to illustrate this argument by focusing on the special case of Szilard's engine in which the central notion is that of a classical measurement of the position of a particle: whether it is on the left- or right-hand side of a box. This set up has been central in the discussion about Maxwell's Demon since Szilard's 1929 paper. Szilard thought that the essential feature of Maxwell's Demon was what he called "intelligence" rather than the complexity of a multi-particle system, and this is why he proposed to study a gas consisting of *one* particle in a box. However, we do not attempt to write down detailed Hamiltonians since the number of details in a specific example is endless. Our aim here is to demonstrate that if anything stands in the way of implementing a Maxwellian Demon it is *not* the principles of classical mechanics.

Consider the experiment described in Figure A1. A particle G is placed in a box of volume V, which is initially thermally isolated (the arrangement is adiabatic up to Stage f; see below). The particle G can be treated as an ideal gas obeying the equation $PV = kT$, where P is the pressure, T is the temperature, and k is Boltzmann's constant. At Stage a a device D that can measure whether G is on the left- or the right-hand side of the box is prepared in a standard ready state S, while G is free to move around in the entire volume of the box. Figure A1 illustrates the two possible evolutions of the experiment, although of course only one of the two evolutions is realized on any given occasion.[1]

---

[1] In Figure A1 we follow the illustration of the experiment given in Bennett (1982), with some significant changes.

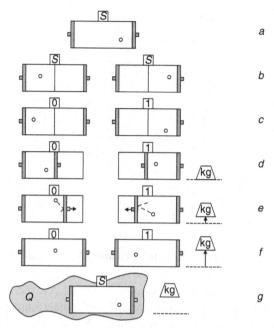

Figure A1 Szilard's engine

The phase-space description of this stage is as follows (see Figure A2). The system consists of three sets of degrees of freedom, two sets belonging to G and one to D. For simplicity, we omit the observer O. The horizontal axis in Figure A2 stands for the position x of G in the box. We ignore the directions y and z since they remain unchanged throughout the experiment. Since, in later parts of the experiments, we shall be interested in whether G is on the right-hand side of the box or the left-hand side of it, we divide the accessible region on x_G into two regions corresponding to the position of G, and denote these regions by L and R.

The axis perpendicular to the page stands for G's velocity v which is determined by the total kinetic energy of G: $E = mv^2/2$. Since the projections of the velocity on its three spatial directions can vary (even when the total kinetic energy E is constant) owing to the collisions with the walls of the box, for any given E the velocity in each spatial direction ranges over a region between $(-v, +v)$ for $v = \sqrt{2E/m}$. Therefore, we represent the velocity macrostate at Stage a as the range v_1 on the axis perpendicular to the page.

The vertical axis corresponds to D's memory state, which is divided into three macrostates $[S]$, $[0]$, and $[1]$. We assume for simplicity that D's

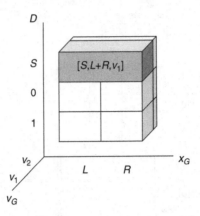

Figure A2 State space of Stage *a*

macrostates are of equal Lebesgue measure. As we shall see later, this assumption is natural but not necessary.

At Stage *a* of the experiment depicted in Figure A1, the actual micro-state of $D + G$ is in the region corresponding to the macrostate $[S, L + R, v_1]$, as illustrated in Figure A2. At this stage, the structure of trajectories is such that given the initial constraints if $D + G$ start in $[S, L + R, v_1]$, its trajectory evolves inside this region.

At Stage *b* of the experiment (see Figure A1) a partition is placed (with negligible investment of work) in the middle of the box so that *G* is trapped either on the left- or on the right-hand side of the box. This means that the trajectories of $D + G$ are now confined to either *L* or *R* and no longer pass between the left- and right-hand sides of the box.

It is important to distinguish, at Stage *b*, between the structure of the trajectories of $D + G$ in the phase space and the macrostate of $D + G$ (see Figure A3). In Figure A3 this fact is expressed as follows: the macrostate $[S, L + R, v_1]$ is depicted by the shaded area, within which there are two disconnected parts of the bundle of trajectories represented by the dashed lines. This means that while the trajectories' structure changes in the transformation from Stage *a* to Stage *b*, the macrostate remains unchanged. This is because whereas at Stage *a* the dynamical structure of the phase space may be topologically *connected*, at Stage *b* the region occupied by $[L + R]$ is necessarily topologically *disconnected*. The classical dynamics does not allow a transformation that makes connected trajec-tories disconnected (or vice versa); this is possible only for projections of trajectories on some subset of the degrees of freedom of the system in question. And so the transformation from Stage *a* to Stage *b* necessarily

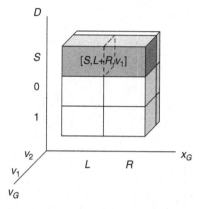

Figure A3 State space of Stage *b*

involves the intervention of some degree of freedom not indicated in Figure A1, such as the partition itself and the automata that manipulate it. In our description of the experiment these additional degrees of freedom are treated as *external constraints*, and they impose limitations on possible evolutions of $D + G$. By distinguishing between the degrees of freedom of $D + G$ and the degrees of freedom of the external constraints, it is possible to account for notions such as preparation, which will turn out to be important later. Despite the change in the structure of the trajectories, the macrostate of $D + G$ at Stage *b* is still, of course, [S, $L + R$, v_1], as in Stage *a*.

In thermodynamics, the transition from Stage *a* to Stage *b* implies that the volume of the box accessible to G is reduced from V to $V/2$. By inserting the partition adiabatically no energy is invested in G and therefore G's velocity is unchanged. Consequently, the pressure on the walls of the box is doubled from p to $2p$. This increase in p is brought about solely by the increase in the frequency of the collisions of G with the walls of the box due to the reduced volume. If we take G to be an ideal gas this entails that G's temperature is unchanged.

The measurement of the location of G by D is described in this experiment by the transition from Stage *b* in Figure A3 to Stage *c* in Figures A4 and A5. The interaction Hamiltonian brings about *correlations* between the macrostates of D and the macrostates of G, such that trajectories that start in [S, L, v_1] end in [0, L, v_1] and trajectories that start in [S, R, v_1] end in [1, R, v_1] (where [0, L, v_1] is the macrostate in which D registers the outcome 0, G is located in the left side of the box, and its velocity is in the range v; and this is also the case for [1, R, v_1]). The dashed lines in

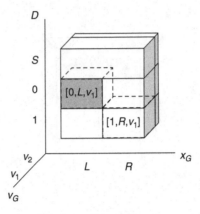

Figure A4 State space of Stage *c*: outcome *L*

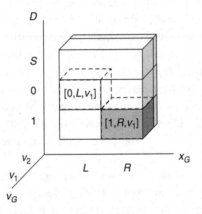

Figure A5 State space of Stage *c*: outcome *R*

Figures A4 and A5 represent the split blob in the measurement and the shaded areas the collapsed outcome (detection and expansion).

The evolution of the blob satisfies *Liouville's theorem* since the Lebesgue measure of the region $[S, L+R, v_1]$ is equal to the Lebesgue measure of the *union* of the regions $[0, L, v_1]$ and $[1, R, v_1]$. In this special case the blob and the two possible macrostates at the end of the measurement exactly overlap (compare Figures 9.3 and 9.4).

The only constraint that follows from Liouville's theorem with respect to the change in the macrostates of $D+G$ during the evolution is that the total Lebesgue measure of the (union of) the macrostates into which the blob evolves cannot be *smaller* than the Lebesgue measure of the blob.

That is all. Thus we can see that our setting satisfies this constraint, since the two macrostates at Stage c exactly overlap with the blob.

This exact overlap, however, is not necessary for the construction of a Demon.

Our account of the experiment so far is different from Szilard's (1929) account (see also Earman and Norton 1998). Szilard argued that in measurement the entropy of $D + G$ decreases, and therefore there *must* be an increase of entropy elsewhere in the universe. We have just shown that the total entropy of $D + G$ may decrease in measurement in a way that is consistent with the classical dynamics, in particular with Liouville's theorem. Therefore, unless one *presupposes* that the entropy of a closed system cannot decrease as stated by the Second Law of Thermodynamics (thus begging the question of Maxwell's Demon) no compensating increase of entropy is required. It seems to us that Szilard was in the right direction in arguing that the entropy of $D + G$ decreases during the measurement (assuming the same partition of the phase space into macro-states). But he failed to realize the full implications of his idea since he did not pursue to the end a purely mechanical analysis.

We now move on to the transition starting at Stage d through Stage e and ending at Stage f of the experiment. In our set up this process is carried out adiabatically. At Stage d we press a piston against vacuum (in accordance with the outcome of the previous measurement), remove the partition without investing any work, and then, at Stage e, release the piston quasi-statically allowing the pressure exerted by G to produce work, until at Stage f the particle G is again free to move throughout the volume V of the box. The work produced by G on the piston is stored in some external degree of freedom, say a weight that is lifted outside the box (see Figure A1). This quasi-static expansion of an ideal gas is the paradigmatic case of a reversible (that is, entropy-conserving) process in thermodynamics. The particle G exerts work on the piston, thereby transferring energy to the weight outside the box. Consequently the free energy of the weight increases, by the amount mgh, where m is the mass of the weight, g the gravitational acceleration, and h the height to which the weight is lifted. Conservation of energy entails that the internal energy of G decreases by an amount equal to the increase of the free energy of the weight. (We focus on an adiabatic process here, but a similar argument can be run by considering an isothermal process; see below.)

Now, it follows from the ideal gas law that the increase in the volume accessible to G (from either L or R to $L + R$) is accompanied by changes both in the pressure and the temperature. The temperature decreases

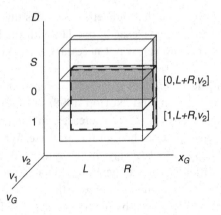

Figure A6 State space of Stages *d* to *f*: outcome *L*

because the temperature of an ideal gas is proportional to its internal energy, which has decreased owing to the transfer of energy from *G* to the weight. The pressure decreases during this transition both because of the decrease in the average kinetic energy (and thus the velocity) of *G* and because of the increase in the distance between the walls of the box.

Note that since the weight here ends at a certain fixed height *h*, which is essentially determined by the momentum it gains from *G*, it can be held in place at its maximal altitude with a negligible investment of work. The mechanism places the weight in a *standard* position at height *h* which does not depend on whether *G* is in [*L*] or [*R*], and then returns to its ready state. Of course, the entire evolution of the mechanism will depend on the memory macrostate of *D*. But the set up is such that once the weight is in its final place and the mechanism is back in its ready state, the only traces of whether *G* was in [*L*] or in [*R*] are in *D*'s memory. We will come back to the issue of *D*'s memory below.

In terms of phase space, the transition from Stage *c* to Stage *f* is as follows (see Figures A6 and A7; for simplicity we do not draw the phase space at the intermediate Stages *d* and *e*). From Stage *c* to Stage *f* there is an increase in the Lebesgue measure of the blob along the x_G axis from either [*L*] or [*R*] to [*L*+*R*], and a *simultaneous* decrease of the measure along the v_G axis from v_1 to $v_2 = \frac{1}{2} v_1$, in such a way that the total Lebesgue measure of the Poincaré section at all times is conserved in accordance with Liouville's theorem. By the end of Stage *f* the macrostate of *D* + *G* is either [0, *L*+*R*, v_2] or [1, *L*+*R*, v_2] as represented by Figures A6 and A7. During this transition the low entropy of *D* + *G* is conserved, while the weight is reversibly lifted. The free energy of the

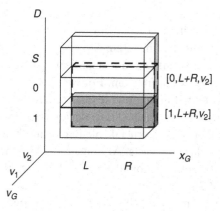

Figure A7 State space of Stages *d* to *f*: outcome *R*

weight can be now used in order to compress another ideal gas G' in a reversible and entropy-conserving way.

Recall that inserting the partition at Stage *b* resulted in the splitting of the phase space into two disconnected regions [*S, L*] and [*S, R*]. However, during the expansion stages from Stage *c* to Stage *f* the gradual changes in the constraints translate into a topological change in the trajectories of *G*, namely that the region [*L + R*] at Stage *f* is connected again. A similar topological change occurs in *D*, so that by the end of Stage *f* the phase space is connected as indicated in Figure A2. Note further that in this transition the entropy of *D + G* is restored to its initial value, i.e. it is doubled. The entropy of *G* is doubled in this transition while the entropy of *D* remains unchanged. The weight remains in its lifted state.

Once again, we have followed the assumption in the literature that the partition of the phase space to macrostates matches exactly the spread of the blob. This assumption is not necessary but it makes the discussion simple, and it has the consequence that the total entropy of *G* (and of course of *D + G*) is conserved during this evolution from Stage *c* to Stage *f*. The crucial point that needs to be stressed here is that there is *no* general argument *based on mechanics* (or statistical mechanics) to the effect that the entropy in *v* must increase during the transition from Stage *c* to Stage *f*, so as to result in an increase of the total entropy of *D + G*.

We now wish to close the operation cycle. We shall say that the cycle is closed if the following conditions are satisfied: (i) *D + G* returns to its initial macrostate; (ii) there are no *macroscopic traces* of the previous history of the universe[2] (e.g. concerning the outcome of the measurement, the memory of D) in either $D + G$ or in the environment Q; (iii) the free energy of the weight at the final state is the same as at Stage f.

We treat the environment Q as a mechanical system which may consist of any number of degrees of freedom. For simplicity, we take Q to consist of one particle. Note that up to now we did not refer to Q since we carried out an adiabatic process in which Q is irrelevant. However, as we shall see now the environment plays a crucial role in closing the cycle.

Here is a way to close the cycle. At Stage f, the blob describing $D + G$'s evolution covers the entire region $[0 + 1, L + R, v_2]$ (see the dashed lines in Figures A6 and A7). We now construct an evolution which maps the points in this blob back to the initial macrostate $[S, L + R, v_1]$ in Figure A2. Here it is important to keep in mind that Figures A6 and A7 describe two possible histories only one of which is the actual history of the universe: each of them describes one out of the two possible macrostates of the universe; but in both figures the entire dynamical blob is depicted. The measure of the blob $[0 + 1, L + R, v_2]$ in Figures A6 and A7 is *equal* to the measure of the final macrostate $[S, L + R, v_1]$ in Figure A2: while the measure of the blob along the D degree of freedom decreases by half, from the union of $[1]$ and $[0]$ to $[S]$, the measure of the blob along the v_G degree of freedom is doubled, since the measure of v_1 is twice the measure of v_2. Therefore, Liouville's theorem is satisfied.

The transition from Stage f to Stage g completes the *erasure* of any traces of the actual position of G at Stage b and any record in D of the outcome of the measurement at the end of Stage c. Since the dynamics is blending in this transition, it is impossible to reconstruct the earlier macrostate of either D or G from the final macrostate of $D + G + Q$ in Stage g, and therefore this is a macroscopic erasure.

The evolution just described in the transition from Stage f to Stage g requires a source of energy, since it involves an increase in the Lebesgue measure of the blob along the v_G degree of freedom corresponding to the increase in G's speed. In other words, in restoring G's initial macrostate, its internal energy increases (by the amount corresponding to the change from v_2 to v_1). Here Q comes into play. Consider Figure A8. Up to now in Figures A2–A7 we have depicted three sets of degrees of freedom. We now need two more, namely the velocity v_Q and position x_Q of Q. We describe them separately in Figure A8, but of course the dimensions of Figure A8

---

[2] Of course, there are always microscopic traces because of the determinism of classical mechanics.

Phase space of the
environment *Q*

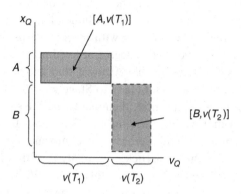

Figure A8 State space of *Q*

belong to the same phase space of Figures A2–A7. This means in particular according to Liouville's theorem that the total volume of the blob describing $D + G + Q$ must be conserved. Since in the transition from Stage f to Stage g the volume of the blob in the $D + G$ degrees of freedom is conserved, and since the total volume of the blob describing $D + G + Q$ must be conserved, it follows that we need to show now that the volume of the blob describing Q alone must be conserved during the interaction of Q with $D + G$. Here is a way to achieve this.

Throughout the Stages a to f, Q is in a fixed macrostate $[A, v(T_1)]$, where A is the position macrostate of Q and $v(T_1)$ is its velocity macrostate corresponding to the initial temperature T_1 of Q. As a result of the interaction of G with Q, energy flows from Q to G. We assume here that all the microstates in the initial macrostate of Q (and, in particular, all the pairs that are the velocity reversal of each other) evolve in such a way that energy is transferred from Q to G.[3] The blob for Q evolves from the region overlapping with the macrostate $[A, v(T_1)]$ to the region overlapping with the macrostate $[B, v(T_2)]$, where T_2 is the temperature of Q at the final state. The final temperature T_2 of Q is lower than its initial temperature T_1 by exactly the amount of energy transferred to G. Therefore, the volume

---

[3] This assumption is not specific to Maxwell's Demon, but is general. For example, in a measurement interaction we assume that all the microstates in M_0 evolve into the union of the macrostates corresponding to the measurement outcomes. Also, this idea is consistent with the time symmetries of mechanics and does not single out a direction in time. A direction of mechanical processes in time can be singled out only by a Past Hypothesis, as we saw in Chapter 10.

of the blob in Figure A8 along the v_Q degree of freedom decreases. Liouville's theorem dictates that (since we required that the total volume of the blob of $D + G$ remain fixed) the volume along x_Q must increase by the same amount. This increase in the volume of x_Q is expressed by the transition to the region overlapping with macrostate $[B, v(T_2)]$ in which the spread of the blob over the x_Q degree of freedom is larger than in the macrostate $[A, v(T_1)]$. Here we have shown that Liouville's theorem is satisfied by the transition from Stage f to Stage g.

If we think of Q as one particle, then increasing the spread of Q along x_Q simply means that we have less information about the position of Q after the interaction (whereas we have more information about Q's velocity). If Q consists of many particles, the analysis of the spread along x_Q is a straightforward generalization of this point. We reiterate that the increase in the spread of the blob here only means a decrease of information about where Q is. In classical mechanics nothing prevents such an increase in x_Q nor are there any mechanical pre-conditions concerning the rest of the universe that need be satisfied in order for this evolution to take place.[4]

Let us consider the changes in the entropy of $D + G + Q$ during the transition from Stage f to Stage g. At the end of the cycle $D + G$ returns to its initial macrostate, and the weight remains lifted. The particle Q evolved from $[A, v(T_1)]$ into $[B, v(T_2)]$. As before, we assume here that the blob pertaining to Q exactly overlaps with the macrostates of Q. Therefore Q's entropy if measured by the Lebesgue measure is conserved. The total entropy of $G + D + Q$ is conserved along the cycle, but the weight has higher free energy.

In general, the question as to the *entropy* of Q during this transition depends on the partition of the phase space of Q into macrostates. Our choice of macrostates in the figures is natural but – as we have stressed throughout the book – is not necessary: the entropy of Q may increase or decrease depending on the partition into macrostates. The macrostate of Q could be larger or smaller in Lebesgue measure, by for example dividing the region overlapping with $[B, v(T_2)]$ into two macrostates. In such a case the spread in x_Q would be exactly the same as in $[A, v(T_1)]$ despite the decrease in the spread along v_Q. The crucial point is that such construction does not seem to violate any known law of mechanics; in particular it

---

[4] In particular, this evolution does not require any pressure difference between Q and the rest of the universe at any time.

obeys Liouville's theorem.[5] Since the macroscopic erasure is relative to a given set of macrostates, the change of entropy during the erasure varies according to the measure of the relevant macrostates. In the case considered here, the transition from Stage *f* to Stage *g* is entropy-conserving. We have here a Maxwellian Demon that is consistent with the laws of statistical mechanics, or so it seems.[6]

Since Szilard's engine as constructed here is a specific model of a Maxwellian Demon, unlike our general state-space argument in Chapter 13, some questions concerning its details remain open. Here are two of them.

(i) The macrostates of *G + D + Q* exactly overlap with the blob during the evolution. It is not clear whether the dynamics and macrostates of thermodynamic systems give rise to such an overlap. At first sight we do not see why not.

(ii) The fact that *G* consists of one particle plays a crucial role in controlling the energy transfer to the weight. But it is an open question whether a single particle is a thermodynamic system, that is, for example whether the ideal gas law applies to it, or whether one particle can be said to have a thermodynamic equilibrium state. A similar question arises with respect to the environment *Q*. It is not clear whether the energy transferred from *Q* to *G* is in the form of heat.[7] Similar questions arise with respect to Maxwell's original Demon.

---

[5] Note that here the environment *Q* is not a heat bath in the thermodynamic sense. It is a consequence of energy conservation that in our set up *Q*'s temperature varies.
[6] See further details in this argument in our 2011 paper (Hemmo and Shenker 2011b).
[7] For example, Jauch and Baron (1972, pp. 171–2) think that Szilard's engine is inadequate for investigating the Second Law because a single particle means that the insertion of a partition confines the gas to half of the container without investing work, thereby violating Gay-Lussac's law. Costa de Beauregard and Tribus (1974, p. 179) correctly reply that such a violation of Gay-Lussac's law is forbidden only because of thermodynamic considerations, while the Demon is meant to inspect thermodynamics.

Appendix B
Quantum mechanics

In Chapter 3 of this book we described some essential features of classical mechanics, which formed the basis for the subsequent discussion throughout the book. In particular, the notions of macrostates and dynamical blob, as we defined and used them, assume classical mechanics, and hence also the Probability Rule in statistical mechanics is essentially classical. As is well known, classical mechanics is no longer taken to be the fundamental physical theory of the world. It has been replaced by quantum mechanics and the theory of relativity. However, it seems to us that the main ideas developed in this book in the classical context translate, *mutatis mutandis* of course, to quantum mechanics. These contain the notions of macrostates and dynamical blobs, and hence also the Probability Rule. Despite the significant difference in the details of these notions, we believe that the main insights gained in this book carry over to discussions based on quantum mechanics.

Now, as is well known, standard quantum mechanics suffers from the notorious measurement problem, which arises because the so-called collapse of the quantum state in measurement and the essential role it gives to the observer are unexplained. The role of the observer in the classical theory, and in particular our notion of the classical collapse, has some striking similarities to von Neumann's projection postulate in the quantum mechanical theory of measurement. However, unlike the situation in quantum mechanics, as we have seen the observer in statistical mechanics is both unavoidable and unproblematic. Still, it is of extreme importance when considering quantum mechanics to bear in mind two points. First, *none* of the points concerning the role of the observer in the classical theory is essentially related to the quantum mechanical measurement problem. Second, the essential *classical* role of the observer in the context of underwriting the thermodynamic regularities will go over in its entirety to quantum statistical mechanics – that is, the observer in quantum statistical mechanics is essential for exactly the same reasons as in the classical theory. Thus the two issues ought to be kept apart.

To see how some of the major issues that arise in classical statistical mechanics also arise *mutatis mutandis* in the context of quantum statistical mechanics, we present here three examples. In Appendix B.1 we address an attempt to underwrite thermodynamics on the basis of quantum mechanics with collapse, specifically the GRW theory, and we point out some characteristics of this approach, in particular the way in which the notions of probability and typicality arise in it. In Appendix B.2 we describe the notions of probability and typicality in Bohmian mechanics which is a completely deterministic hidden variables theory with no collapse of the quantum mechanical wavefunction. And in Appendix B.3 we construct a quantum mechanical Maxwellian Demon, in both collapse and no collapse quantum mechanical theories (with no extra variables).

B.1 Albert's approach

An attempt at underwriting the thermodynamic regularities (and the standard probabilities in classical statistical mechanical) by the dynamics in quantum mechanical collapse theories (of a certain kind; see below) was made about 10 years ago by David Albert.[1] Albert's approach has some extremely interesting implications (which are unavailable in the classical context) concerning the intertwined notions of probability and typicality in statistical mechanics on which we shall focus here.

Consider a thermodynamic system (a gas, say) that is prepared in a low-entropy macrostate $[M_0]$; for example, $[M_0]$ may correspond to a gas being confined by a partition to the left half of a box. We know from thermodynamics that if we lift the partition at t_0 and if the gas is not subsequently disturbed, then its entropy will increase such that after some time Δt the gas will fill the entire volume of the box. We saw how statistical mechanics can account for this behavior in terms of the Probability Rule: that is, roughly, we get high probability for higher-entropy macrostates M_i at some later time t_1 if and only if the regions of overlap between the evolved dynamical blob $B(t_1)$ and the high-entropy macrostates M_i have high μ-measure (see Chapter 7 for more details).

---

[1] Albert (2000, Ch. 7). See also Albert (2012).

According to Albert, the structure of statistical mechanics in the classical context that best explains thermodynamic behavior depends (as in our approach) on the interplay between macrostates and dynamics, but this interplay is somewhat different from the one put forward in this book. There is a fact about the way in which the state space is partitioned into thermodynamic macrostates (as we saw in Chapter 5, this fact depends on our physiological structure as observers). And the dynamics, which is taken to be given by the classical equations of motion, satisfies three conditions: the Statistical Postulate (for details see Chapter 6), the Past Hypothesis (for details see Chapter 10), and a certain dynamical hypothesis, which we now describe briefly.

The dynamical hypothesis that Albert puts forward concerns the *structure of trajectories* of thermodynamic systems, and is a conjunction of several assumptions as follows. The microstates in the initial macrostate are supposed to be divided into two sets: those called "normal," which lie on trajectories that give rise to macroscopic evolutions that satisfy the laws of thermodynamics, and the "abnormal" ones, which do not give rise to evolutions that satisfy the laws of thermodynamics. Note that this is a dynamical assumption since it is about the regions in the phase space through which trajectories pass. Suppose now that it is in fact true that the Lebesgue measure of the normal set is very high; and suppose that this measure corresponds to *probability* (as in the typicality approach; see below), so that the probability that a randomly sampled microstate from the initial macrostate is normal is very high. And suppose further that the thermodynamic abnormal microstates are distributed more or less uniformly amongst the normal ones. In other words, suppose not only that the Lebesgue measure counts the abnormal microstates in $[M_0]$ to be rare but also that the abnormal microstates are distributed (or scattered) as it were over the normal ones more or less uniformly over the region $[M_0]$.

This dynamical hypothesis (taking all the above assumptions together) entails thermodynamic behavior with high probability for all possible macrostates in the phase space of a system. But note that it states contingent matters of fact: for example, the hypothesis is false in a universe in which there is a Maxwellian Demon (see Chapter 13), and true in a universe in which there is no Demon; and since nothing in mechanics prohibits Demons, whether our universe is of the first kind or of the second kind is a matter of fact to be discovered empirically. In this sense, Albert's dynamical hypothesis is an empirical one.

As we saw in Chapter 6, Albert's approach to classical statistical mechanics (relying on his Statistical Postulate) can be understood in two ways: either as belonging to the typicality approach or to the approach put forward in this book. In the typicality approach Albert's Statistical Postulate is translated into the idea that the Lebesgue measure has a preferred status of a natural measure which gives rise to a natural probability distribution over the initial macrostates; the reason for this preferred status may, for example, be that it is preserved under the dynamics by Liouville's theorem. And it so happens that on this measure the set of the normal microstates is overwhelmingly larger than the set of the abnormal microstates. And it is this difference in the measure of the normal and abnormal sets that is supposed to explain our thermodynamic experience. (This is the explanatory order according to the typicality approach, discussed in Chapter 8.) Alternatively, one can take the choice of the Lebesgue measure in Albert's approach to be the result of a mere simple generalization of our finite experience of the relative frequencies of transitions between macrostates, as outlined in Chapter 6 above.

Whether one understands probability in Albert's Statistical Postulate in classical statistical mechanics in term of typicality, or as resulting from the Probability Rule put forward in this book, the following question arises. Can one derive the statistical mechanical probabilities (either the Statistical Postulate or the Probability Rule) from the quantum mechanical probabilities and thus remain with only one kind of probability in physics? The classical probabilities seem to have nothing to do with the probabilities that we encounter in quantum mechanics. This is because the classical probabilities are introduced as probabilities about the *macroscopic* evolution of a thermodynamic system. And therefore it seems that they will have to be introduced over and above the quantum mechanical probabilities. Albert's essential idea is that one can get rid of the classical probabilities altogether by replacing the classical dynamics with the right sort of quantum mechanical dynamics and still be able to explain thermodynamics along the lines sketched above. Let's see how this is supposed to work.

Consider again Albert's dynamical hypothesis in the classical context, in more detail. Suppose to start with that the set of all the normal microstates in our initial macrostate (say $[M_0]$) is vastly larger (by the Lebesgue measure) than the set of all the abnormal microstates, not only over the entire phase space region covered by $[M_0]$ but also in every *small enough* subregion of $[M_0]$. And moreover, suppose that this happens to be true for every small enough subregion in the neighborhood of *every single* abnormal microstates in $[M_0]$.

Albert's idea is that given this dynamical condition, almost any small perturbation on the system will send it to a normal microstate. In other words, if the structure of trajectories of thermodynamic systems satisfies this dynamical condition, then the property of being a normal microstate is highly stable under minute external perturbations (assuming that the perturbations are not correlated with the abnormal microstates in some conspiratorial way). This means that if we have a source of *random* perturbations, and if these perturbations would result in the right way from the dynamics, then we will have no need whatsoever for the classical probabilities in order to obtain a probabilistic underwriting of thermodynamics. Albert's idea is that the quantum mechanical collapses of the wavefunction in theories of spontaneous collapses like the theory proposed by Ghirardi, Rimini, and Weber[2] (GRW henceforth) can serve as precisely the source of the random perturbations we need.

Of course the idea that the property of being a thermodynamic normal microstate is stable under random perturbations essentially stands behind the so-called interventionist approaches also in classical statistical mechanics.[3] But in the classical context there are no genuinely random perturbations, and so one needs to introduce some sort of a probability distribution over the source of the perturbations, say over the microstates of the environment, in order to complete the picture. But in the context of the GRW theory the source of the perturbations just is the quantum mechanical collapses of the wavefunction, which are genuinely random since the probabilities come from the stochastic dynamics of the wavefunction. And so if we take the GRW theory as our fundamental dynamical theory, then we can let go of the classical probabilities in statistical mechanics since they would have no work to do. This result is, of course, desirable in the sense that it is elegant and ontologically parsimonious.

To see how this can be made to work, we only need to understand the way in which the GRW theory cooks up the collapse of the quantum mechanical wavefunction. Ghirardi, Rimini, and Weber suppose that the dynamics of the wavefunction of any single particle system is almost always correctly described by the Schrödinger equation (in the non-relativistic case). But once in every 10^{10} years or so (that is, at random but with fixed probability per unit of time) the wavefunction *jumps*, or gets localized in space. Mathematically this is represented by multiplying the wavefunction by a narrow shaped Gaussian in position of width equal to about 10^{-5} cm (which is roughly the diameter of one of the lighter atoms).

---

[2] Ghirardi, Rimini, and Weber (1986).
[3] See for example Bergmann and Lebowitz (1955), Earman and Redei (1996), and Ridderbos and Redhead (1998).

The probability that this Gaussian is centered around any particular spatial point x is taken to be equal to the modulus square of the amplitude of the quantum mechanical wavefunction at that point just before this jump. It has been suggested by John Bell[4] that such a GRW jump may be understood as the occurrence of a *flash* at the spacetime point x and that such flashes exhaust the entirety of what it means for something to exist in spacetime.[5] The probability per unit time for a GRW jump and the width of the GRW Gaussian are taken as new constants of nature. This is the entire theory. It can be shown that all the empirically confirmed predictions of the standard quantum mechanical algorithm (including characteristic quantum mechanical behavior as well as the classical-like behavior of macroscopic systems) can be derived from this theory.

And now Albert's idea is that if the structure of the trajectories in the state space conforms more or less to the assumptions above, then the GRW jumps are precisely the sort of random perturbations we need in order to prove that the thermodynamic behavior of systems like our gas is overwhelmingly likely, regardless of the initial conditions. Here is how the idea is meant to work in a bit more detail. Suppose that the wavefunction of a gas immediately after a GRW jump at some point x is given by $\psi(x, t)$. This wavefunction gives, according to the GRW theory, for any time t after the first jump, an infinite set of probability distributions $P_{\psi(x, t)}(x, t)$ over all the possible spacetime points at which the *next* jump may occur given the wavefunction just before this next jump. And now to get thermodynamics out of this all we need is that: (I) the rate of the GRW jumps is high relative to the rate of characteristic thermodynamic evolutions; (II) the size of the GRW Gaussian is such that a flash occurring at *any point* x is highly likely to be on a normal point for any possible wavefunction of the gas before the flash. This is once again Albert's dynamical hypothesis and it must hold for any conceivable wavefunction of the gas, if no probability distribution other than $P_{\psi(x, t)}(x, t)$ is to be introduced. Since the probabilities for GRW jumps do not depend in any other way on the previous history of the gas, it follows (if these two conditions are satisfied) that at all times it is extremely unlikely that the GRW flashes will occur at abnormal points, independently of initial conditions. In particular the small probability for a thermodynamic

---

[4] Bell (1987).
[5] Bell's idea has been further developed by Tumulka (2006) who was able to write down an explicitly *relativistic* model of non-interacting particles in the GRW theory – under the flash interpretation.

abnormal behavior is independent of whether or not the post-jumps wavefunctions at any previous time in the history of the gas was abnormal.

It is interesting to consider the implications of Albert's approach in some special cases such as very small gases in which (in the GRW theory) the probability for a flash is small, or the spin echo experiments, in which the atoms almost do not move in space. But in all these cases it turns out that Albert's proposal plausibly works given the spontaneous character of the GRW jumps and the supposed dynamical hypothesis.[6]

Albert's approach implies (but does *not* assume) that the quantum mechanical measure over the initial macrostate can be understood as probability. In his approach the mechanism of the GRW jumps plays the role of randomly (relative to the quantum mechanical measure) sampling a normal point in $[M_0]$ with high probability via $P_{\psi(x,t)}(x,t)$. And this is precisely what the typicality approach needs in order to make a linkage between measure and probability. By putting the seat of probability in the GRW dynamics Albert gets not only a single origin for probability in physics but also a straightforward translation of *non-probabilistic* mathematical facts about measure in the classical theory into facts about physical probability in the quantum theory. This strategy is available only in theories like the GRW theory in which the dynamics of the quantum mechanical wavefunction is genuinely stochastic in a way that guarantees that the probabilities for the perturbations themselves are lawlike and *cannot* be circumvented.

However, two points need to be stressed here. Let us suppose that the GRW theory is the true dynamical theory of the world. In order for Albert's approach to work such that one need not introduce any facts about probability over and above those given by the GRW dynamics, it must be true that the structure of trajectories in the state space of every thermodynamic system is such that *every* initial wavefunction of the system that evolves under the GRW dynamics is highly likely to collapse at every moment onto a thermodynamically *normal* quantum mechanical wavefunction. If this condition will not hold universally, that is, if this condition fails to be true even only with respect to a single initial wavefunction of the system (call it a *deviant* wavefunction), the GRW dynamics will not be enough to deliver the right statistical mechanical probabilities. In this case one would have to resort to the standard

[6] For more details see Albert (2000, Ch. 7) and also our previous work (Hemmo and Shenker 2001, 2003, 2005).

approach and add over and above the GRW dynamics a classical probability measure over the space of wavefunctions that would yield that such deviant wavefunctions should be somehow avoided with high probability.

The second point refers to the issue of Maxwell's Demon in the quantum mechanical context. In Appendix B.3 we shall see that as a matter of principle Maxwellian Demons can be cooked up even if one replaces the classical equations of motion with the quantum mechanical ones, and in particular with the GRW dynamics. Of course in the GRW theory Maxwellian Demons cannot be taken to control the probabilities for the perturbations since these are lawlike by assumption. And so the fact that Demons are consistent with the GRW theory entails that Albert's non-probabilistic assumption concerning the dynamical hypothesis about the structure of trajectories is not always true of quantum mechanical systems. And therefore as a matter of fact it might be true that for some initial wavefunctions of the universe (given some appropriate dynamical set up) the probability for GRW collapses onto deviant wavefunctions are *not* small.[7]

B.2 Bohmian mechanics

In Chapter 8 we distinguished between two readings of typicality, *a priori* and *a posteriori*. In statistical mechanics the a posteriori reading is based on the Probability Rule in which the choice of measure is based on experience. The *a priori* reading, by contrast, attempts to justify the choice of measure on other grounds based on veraious properites of the measure, for example, conservation under the dynamics. We argued that the *a priori* reading is unsound while the *a posteriori* reading is (inductively) sound but cannot be taken to explain our experience, on which it is based. In this appendix we address a similar situation in the context of Bohm's theory.

---

[7] It is important to note in this context that the standard models of environmental decoherence in quantum mechanics *presuppose* a probability distribution over the states of the environment in order to show that it is highly probable (given the interaction Hamiltonian) that the states of the environment that get coupled to different (but superposed) states of the system are (approximately) orthogonal. This probability distribution is *not* derived from the quantum mechanical transition probabilities given by the Born rule. And so this probability distribution plays the role of the probability distribution over initial conditions in classical statistical mechanics.

Dürr, Goldstein, and Zanghi[8] (hereafter DGZ) appeal to a typicality argument (of the *a priori* kind) in order to explain why Bohm's theory recovers the quantum mechanical probabilities (as given by Born's rule). DGZ prove that:

> ... To demonstrate the compatibility of Bohmian mechanics with the predictions of the quantum formalism, we must show that for at least some choice of initial universal Ψ and q, the evolution [given by Bohm's velocity equation] leads to apparently random pattern of events, with empirical distribution given by the quantum formalism. In fact we show much more. We prove that for every initial ψ this agreement with the predictions of the quantum formalism is obtained for typical – i.e. for the overwhelming majority of – choices of initial q. And the sense of typicality here is with respect to the only mathematically natural – because equivariant – candidate at hand, namely, quantum equilibrium. Thus, on the universal level, the physical significance of quantum equilibrium is as a measure of typicality, and the ultimate justification of the quantum equilibrium hypothesis is, as we shall show, in terms of the statistical behavior arising from a typical initial configuration.[9]

In other words, DGZ argue that for every initial universal wavefunction and for *a typical* initial global configuration, the probability distribution over the Bohmian position of subsystems of the universe is given by the absolute square of the effective wavefunction of the subsystems (when an effective wavefunction exists). From this result they show that the conditions sufficient for the laws of large numbers to hold are satisfied in Bohmian mechanics with probability distribution that recovers the predictions of standard quantum mechanics as given by Born's rule. Here the notion of a typical global configuration is understood relative to the quantum mechanical measure, that is, the absolute square of the *universal* wavefunction.

What is the role of the typicality assumption in this argument? Typicality is meant to replace here probability over initial conditions of the universe, and thus explain the initial conditions while avoiding what Albert[10] calls the fairy tale concerning the random or probabilistic *choice* of initial conditions. As we argued in Chapter 8, there are two problems with this approach. First, if the notion of typicality is non-probabilistic, it is unclear why a condition that is true for most initial conditions (relative to the measure) should be taken as true for a given system. The problem is

---

[8] See Dürr, Goldstein, and Zanghi (1992).
[9] DGZ (1992), p. 859.
[10] Albert (1992, Ch. 7).

to justify the choice of the measure of typicality (in this case, the absolute square of the universal wavefunction) in a non-circular way. The argument given by DGZ for preferring the quantum mechanical measure as natural is that it is the only equivariant measure under the dynamics of the wavefunction and the Bohmian dynamics of the initial q. However, this argument is irrelevant just as the classical Liouville's theorem is irrelevant as an argument for preferring the Lebesgue measure in the case of ergodic dynamics. Second, we have no *empirical* access to *counterfactual* initial configurations of the universe.

It is important to stress that this criticism concerning the notion of typicality does not in anyway undermine the important results by DGZ concerning the probabilistic content of Bohm's theory. In fact, it seems to us that one can retain all these results by making the simpler assumption that the initial condition of the universe is *as a matter of fact* of the kind that yields the quantum mechanical predictions. This fact need not be further justified; in particular, it need not (and in our view cannot) be justified by reference to a set of typical initial conditions. Instead, we can construct the set of initial conditions and the wavefunction in a way that yields the probabilistic content of the theory, as it is given by the quantum mechanical Born rule (the absolute square of the (effective) wavefunction of subsystems of the universe in Bohm's theory), where the Born rule itself is subject to empirical tests in our experience.

This is enough in order to *describe* our experience in terms of Bohmian Mechanics. Note that here we do not appeal to typicality, and indeed, as we argued in Chapter 8, we do not think that typicality considerations in physics are explanatory. It is tempting to rephrase the account given in the previous paragraph in terms of typicality by finding a measure of typicality according to which the right sort of initial universal conditions turns out to be typical. But evidently, this would be circular reasoning: since the measure of typicality is constructed so as to fit our experience, it cannot be taken to explain experience.

B.3 A quantum mechanical Maxwellian Demon

We now wish to show that according to quantum mechanics Szilard's set up of a particle in a box (described in Figure A.1) can be a Maxwellian Demon. In the foregoing analysis we shall describe the microstate of the entire set up as it is given by quantum mechanics. We do not describe here the macrostates in the set up, since the notion of a macrostate in quantum mechanics is essentially classical, and therefore our argument for a Maxwellian Demon in the classical context would apply *mutatis mutandis* to such a macroscopic description. Our aim here is to show that the quantum mechanical dynamics, and in particular the quantum probabilities do not stand in the way of realizing a Maxwellian Demon.[11]

We start by assuming a quantum mechanical dynamics with no collapses, i.e. dynamics that satisfies the Schrödinger equation at all times. We will subsequently consider a collapse dynamics.

At t_0 the quantum state of the entire set up is the following:

$$|\Psi(0)\rangle = |x_0\rangle_p |S\rangle_m |\text{down}\rangle_w |e_0\rangle_e, \qquad (B.1)$$

where $|x_0\rangle_p$ is the initial state of the particle in a one-dimensional box; $|S\rangle_m$ is the initial standard (ready) state of the measuring device; $|\text{down}\rangle_w$ is the initial state of the weight which is positioned at some initial height we denote by *down*; and $|e_0\rangle_e$ is the initial state of the environment which we assume does not interact with the particle or the weight (or the partition). In particular, this means that the quantum state of the particle and the partition do not undergo a decoherence interaction with the environment.

The initial energy state of the particle $|x_0\rangle_p$ is some superposition of energy eigenstates which depend on the width a of the box where the amplitudes in $|x_0\rangle_p$ give a definite expectation value $\langle x_0 \rangle$ for the energy of the particle. We assume for simplicity that all the energy eigenstates at the initial time have a node in the middle of the box, so that the quantum mechanical probability of finding the particle exactly in the middle of the box is zero.[12] For example, the quantum state of the particle may be the eigenstate of the energy:

$$|x_0\rangle_p = \sin\frac{n\pi x}{a} e^{-(E_n/h)t},$$

with an eigenvalue

$$E_n = \frac{n^2 h^2}{8 m_p a^2},$$

where n is an even number and m_p is the mass of the particle. In the more general case, $|x_0\rangle_p$ will be a superposition of such energy eigenstates. We

---

[11] For a standard analysis of the Demon in quantum mechanics, see for example Zurek (1990).
[12] See for example French and Taylor 1978, pp. 113–17.

reiterate that such a superposition cannot be interpreted as expressing ignorance over the energy eigenstates of the particle. It is the fine-grained quantum mechanical *microstate* of the particle and not a macrostate in any standard sense.

The state $|S\rangle_m$ is the standard ready state of the measuring device which is an eigenstate of the so-called *pointer observable* in this set up. For simplicity we take a spin-1 particle to represent the measuring device with $|S\rangle_m = |-1\rangle$ in the z-direction. The two other spin-1 eigenstates $|0\rangle$ and $|1\rangle$ in the z-direction will correspond to the two possible outcomes of the measurement.

At t_1 we insert a partition at the middle of the box where the wavefunction is zero such that the expectation value for the energy remains completely unaltered. This is of course highly idealized and creates many technical questions. For example, given the quantum mechanical uncertainty relations one cannot insert the partition exactly at a point since in this case its momentum will be infinite, and consequently the partition would be subject to uncontrollable fluctuations. To avoid this, one may assume that the partition and consequently also the particle are relatively massive, but that nonetheless they are kept in isolation from the environment. Alternatively, if there is a certain amount of decoherence, we may assume that the degrees of freedom in the environment are controllable in the subsequent stages of our experiment. Again, we are not concerned here with the question of the feasibility of the experiment as with the question of the consistency of the experiment with thermodynamics.

We therefore take it that ideally at t_1 immediately after the insertion of the partition at the middle of the box the quantum state of the set up becomes:

$$|\Psi(1)\rangle = |x_1\rangle_p |S\rangle_m |\text{down}\rangle_w |e_0\rangle_e, \qquad (B.2)$$

but the expectation value of the energy of the particle is unaltered $\langle x_0 \rangle = \langle x_1 \rangle$.[13] Although the particle's expectation value for the energy does not change in this interaction, the wavefunction of the particle now becomes a superposition of two components

$$|x_1\rangle_p = \frac{1}{\sqrt{2}}(|L\rangle_p + |R\rangle_p), \qquad (B.3)$$

---

[13] In each side of the partition the superposition is of both odd and even eigenstates but the width of the box is now $a/2$.

where $|L\rangle_p$ and $|R\rangle_p$ are (in general) superpositions of energy states of a particle in a box of width $a/2$ with *any* n, where for $|L\rangle_p$ the position x varies over the range $[0, a/2]$ and for $|R\rangle_p$ x varies over the range $[a/2, a]$, and the equal amplitudes are a consequence of our idealization that the partition is placed exactly at the middle of the box.

At t_2 a measurement of the coarse-grained position of the particle is carried out corresponding to whether the particle is located in the left- or right-hand side of the box. The quantum state immediately after the measurement as described by the Schrödinger equation is:

$$|\Psi(2)\rangle = \frac{1}{\sqrt{2}}(|L\rangle_p|\text{left}\rangle_m + |R\rangle_p|\text{right}\rangle_m)|\text{down}\rangle_w|e_0\rangle_e, \qquad (\text{B.4})$$

where $|\text{left}\rangle_m$ and $|\text{right}\rangle_m$ are some *almost* orthogonal states of the device that are close in inner product to, respectively, the pointer states $|0\rangle$ and $|1\rangle$ and are one-to-one correlated (respectively) with the locations of the particle given by the energy states $|L\rangle_p$ and $|R\rangle_p$. But the overall quantum state of the set up is superposed, that is, there is no collapse in the measurement.

The measurement at time t_2 increases the von Neumann entropy of the device m and of the particle (since both enter a mixed state). But this is irrelevant for the question of Maxwell's Demon for two reasons. First, the von Neumann entropy does not correspond to thermodynamic entropy.[14] Second, as we will see, the particle and the device at time t_5 will return to pure states in which the von Neumann entropy decreases to its initial zero value.

At t_3 the partition is replaced by the piston (in both components of the superposition $|\Psi(2)\rangle$), which is coupled to the weight. But we assume for simplicity that the quantum state of the set up remains the same $|\Psi(2)\rangle = |\Psi(3)\rangle$. Note that here we consider the partition and the piston as external constraints, as usual in the literature. Taking them as parts of the system would create technical questions, which are solvable.

At t_4 the piston is pushed (to the left or to the right) by the particle. The quantum state of the set up at t_4 has the form:

$$|\Psi(4)\rangle = \frac{1}{\sqrt{2}}\Big(|L(w(t))\rangle_p|\text{left}\rangle_m + |R(u(t))\rangle_p|\text{right}\rangle_m\Big)|y(t)\rangle_w|e_0\rangle_e, \qquad (\text{B.5})$$

where the energy states $|L(w(t))\rangle$ and $|R(u(t))\rangle$ change with the changing width of the box $w(t)$ and $u(t)$. The expectation value of the particle's energy $\langle x(t)\rangle_p$ is given by

[14] See Hemmo and Shenker (2006).

$$\langle x(t)\rangle = \langle \frac{1}{2}(L(w(t)) + R(u(t)))\rangle$$

and decreases continuously over time in the time interval from t_4 up to t_6. The energy state of the weight $|y(t)\rangle_w$ changes accordingly so that y ranges from the position *down* to the position *up*, and the expectation value of the energy of the weight increases accordingly. Conservation of energy, which in quantum mechanics according to Ehrenfest's theorem is stated in terms of expectation values, implies that

$$\langle x(t)\rangle_p + \langle y(t)\rangle_w = \langle x_0\rangle_p. \tag{B.6}$$

Since the weight is positioned in a gravitational field, higher energy expectation values for the weight are coupled to higher positions of the weight. This means that the quantum mechanical transfer of energy increases the probability that upon measurement the weight will be found at higher positions.

During the expansion of the particle as described by (B.5) the correlations between the memory states of the device and the energy states of the particle are gradually lost. The energy states $|L(w(t))\rangle_p$ and $|R(u(t))\rangle_p$ are such that at any time t the probabilities for finding the particle in the left- or right-hand side of the box are invariably ½. The conditional probabilities, however, for finding the particle in the right- (left-)hand side given that the device is in the state $|\text{left}\rangle_m$ ($|\text{right}\rangle_m$) at t_3 is almost zero, but it increases with time and becomes ½ at t_5 since the correlations between the memory state of the device and the location of the particle are gradually lost. At t_5 the quantum state of the set up is given by:

$$|\Psi(5)\rangle = |x_5\rangle_p \frac{1}{\sqrt{2}}(|\text{left}\rangle_m + |\text{right}\rangle_m)|\text{up}\rangle_w|e_0\rangle_e, \tag{B.7}$$

where $|x_5\rangle_p$ is the energy state of a particle in a one-dimensional box of width a with amplitudes which match the decrease in the expectation value of the energy, $\langle x_5\rangle_p = \langle x_0\rangle_p - \langle \text{up}\rangle_w$, and $|\text{up}\rangle_w$ corresponds to the up-state of the weight corresponding to a higher expectation value of the energy. The measuring device remains in the superposed state $\frac{1}{\sqrt{2}}(|\text{left}\rangle_m + |\text{right}\rangle_m)$, but the particle and the device are now decoupled (and disentangled).[15]

---

[15] Since the energy states of the particle are superpositions rather than classical mixtures, no question of ignorance arises here, and moreover there is no difference in the details of the evolutions of the two components of (B.5). i.e. they are symmetrical.

To complete the cycle of operation[16] we need to return the particle to its initial energy state and erase the memory of the measuring device. Note that closing the cycle does not mean also returning the environment into its initial state. What we need is to erase any trace in the universe from which one could read off the outcome of the measurement. But this has been done already by the transition to the wavefunction in (B.7). In this state there are no traces of the outcome of the measurement in any degree of freedom of the universe. We can now also return the quantum state of the device $\frac{1}{\sqrt{2}}(|\text{left}\rangle_m + |\text{right}\rangle_m)$ to its initial state $|S\rangle_m$. Since these two states are non-orthogonal pure states, there is a quantum mechanical Hamiltonian that will do that.

We believe that by this we have established our task. In particular, as can be seen in equation (B.6) we have transferred energy from the particle to the weight without further changes in the universe.

If we also wish to return the particle to its initial energy state we need of course to take energy from the environment e. For this we let the particle interact with e such that with *certainty* the particle will return to its initial energy state $|x_0\rangle_p$. Since we assume that the environment is initially in a pure state, plausibly very complex but nevertheless a pure state rather than a mixture, there is a deterministic evolution that will yield precisely this energy transfer, no matter how complicated this may be. In this interaction the expectation value of the energy of e decreases by the same amount in which the expectation value of the energy of the particle (or ultimately the weight) increases, $\langle e_0 \rangle_e - \langle e_6 \rangle_e = \langle \text{up} \rangle_w - \langle \text{down} \rangle_w$ as required by conservation of energy. Of course, also the reversed evolution is possible, i.e. that the particle will transfer energy to the environment, but we assume that the initial conditions in our set up match the first course of evolution. The weight's energy expectation value is higher than in its initial state. The cycle is completed and the set up is ready to go once again.

Let us now consider a *collapse* dynamics in the framework of standard quantum mechanics, say in von Neumann's formulation, where now we can even assume that the measurement brings about a collapse of the wavefunction onto exactly the pointer basis states $|0\rangle$ and $|1\rangle$. The quantum state at t_2 just after the collapse is therefore either

$$|\Psi_c(2)\rangle = |L\rangle_p |0\rangle_m |\text{down}\rangle_w |e_0\rangle_e \qquad (B.8)$$

or

---

[16] Albert (2000, Ch. 5) argues that closing the cycle of operation is irrelevant to the question of the Demon. We show in Chapter 13 (see also our 2010 paper, Hemmo and Shenker 2010) how the cycle of operation can be closed in the right sense, but we don't require that the environment will return to its initial state (see below).

$$|\Psi_c(2)\rangle = |R\rangle_p |1\rangle_m |\text{down}\rangle_w |e_0\rangle_e$$

with probability $\frac{1}{2}$, and at t_5 (before the interaction of the particle with e) it is either

$$|\Psi_c(5)\rangle = |x_5\rangle_p |0\rangle_m |\text{up}\rangle_w |e_0\rangle_e$$

or   (B.9)

$$|\Psi_c(5)\rangle = |x_5\rangle_p |1\rangle_m |\text{up}\rangle_w |e_0\rangle_e.$$

The unitary Schrödinger equation cannot map the final states $|0\rangle_m$ and $|1\rangle_m$ of the device m to $|S\rangle_m$. But we can erase the memory of the device m non-unitarily in von Neumann's way[17] by measuring on m at t_6 an observable λ which is incompatible (in the above sense) with the pointer observable. Subsequently, to bring the device back into its standard state, we measure on m the pointer observable until the outcome corresponding to $|S\rangle_m$ is obtained, at which case the cycle is completed. Note that the erasure of the memory does not require additional degrees of freedom in the environment of the set up, and in particular involves no environmental dissipation.

Alternatively, we can map at t_6 the quantum state of the composite system $m + e$ as follows:

$$|0\rangle_m |e_0\rangle_e \rightarrow |S\rangle_m |e_1\rangle_e$$
                                                                                                    (B.10)
$$|1\rangle_m |e_0\rangle_e \rightarrow |S\rangle_m |e_2\rangle_e$$

where $|e_1\rangle_e$ and $|e_2\rangle_e$ are eigenstates of some observable Q of e and the arrow represents the Schrödinger evolution. Note that since the evolution in (B.10) is deterministic the final states are still correlated with the device's memory states, and so at t_6 there are in fact memories in e of the outcome of the measurement at t_2. But we can now measure on e an observable Q' which is *maximally* incompatible with Q in the sense that the quantum mechanical probability distribution (as given by the modulus square of the amplitudes) of the different values of Q given any eigenstate of Q' is uniform. This means that the measurement of Q' on the device m *erases* with certainty any traces in the environment e of the memory state of the device at any earlier time (see the next section).[18]

---

[17] See Ch. 5 of von Neumann (1932).
[18] If we settle for a probabilistic notion of an erasure a measurement of any observable of the environment which does not commute with Q would be enough.

Measurement and erasure

In the literature the discussion on Maxwell's Demon focuses on measurement and erasure. Unlike the classical case, in which erasure is in some sense the reversal of measurement, in quantum mechanics these two notions are intertwined. Consider first measurement and erasure in *no collapse* quantum mechanical theories (i.e. in which the dynamics of the wavefunction is given by the Schrödinger equation). As we saw, the measurement and erasure evolutions of the measuring device m are given by the following sequence:

$$|S\rangle \to \frac{1}{2}\left(|\text{left}\rangle\langle\text{left}| + |\text{right}\rangle\langle\text{right}|\right) \to \frac{1}{\sqrt{2}}\left(|\text{left}\rangle + |\text{right}\rangle\right) \to |S\rangle,$$

$$(\text{B.11})$$

where the first arrow stands for the measurement and the second for the erasure. Since the evolution results in no collapse of the wavefunction after the measurement, the measurement transforms the device m from a pure to a mixed (i.e. *improper*) state, and the erasure transforms m (in the reversed direction) from the mixed to a pure state. Of course, the evolution described by (B.11) is possible since it is brought about by the Schrödinger evolution of the *global* wavefunction $|\Psi(t)\rangle$, which is time-reversible.

It might seem that the second arrow in (B.11) does not in fact erase the outcome of the measurement since the Schrödinger evolution is reversible. But this is mistaken, for the following reasons. First, if one takes it that the second arrow in evolution (B.11) does not bring about an erasure just because the dynamics is reversible, then by the *same* argument, the determinism of the dynamics entails that the first arrow in (B.11) does not in fact bring about a measurement with a definite outcome. Second, as a matter of fact according to all the quantum mechanical theories in which the quantum state does not collapse (e.g. Bohm's theory, Everett's theory) the measurement interaction (represented by the first arrow in (B.11) has in fact an outcome, and the erasure represented by the second arrow in (B.11) in fact erases that outcome. That is, in these theories the outcome of the measurement cannot be recovered once the erasure is completed.[19]

---

[19] We skip the details of how this is accounted for in Bohm's and Everett's theories. But just note that in Bohm's theory there is no *micro*scopic erasure because of the determinism of the guidance equation for the trajectories. In Everett-style theories there might be a microscopic erasure depending on what one says about the evolution of memories when the branches of (B.11) re-interfere.

In standard quantum mechanics with collapse in measurement the notions of measurement and erasure are two aspects of the same physical process. We can represent these two evolutions as follows.

$$|S\rangle \rightarrow \frac{1}{2}\left(|0\rangle\langle0| + |1\rangle\langle1|\right) \rightarrow \frac{1}{\sqrt{2}}(|0\rangle + |1\rangle) \rightarrow |S\rangle. \qquad (B.12)$$

As we saw above, the measurement of the left or right location of the particle is brought about by coupling the memory states $|0\rangle$ and $|1\rangle$ of the device m to the states of the particle, and then collapsing the memory onto one of them. The erasure of the memory of the outcome of the measurement is brought about in the standard theory by *measuring* the observable λ of the measuring device, which is quantum mechanically incompatible with the pointer (or memory) observable (or alternatively by measuring Q' of the environment).[20]

In a sense, in standard quantum mechanics (with or without collapse) a measurement of any observable that does not commute with the memory observable is an erasure, since after such a measurement it is impossible to retrodict the pre-erasure state of the memory system from the post-erasure state. In a collapse theory the situation is even worse: after a quantum erasure one cannot even retrodict which observables took definite values in the past. In our set up, since λ is maximally incompatible with the pointer (or memory) observable, even if we assume that there is a memory of the *identity* of the pointer observable, one cannot retrodict even *probabilistically* which memory state $|0\rangle$ or $|1\rangle$ and therefore which outcome of the measurement took place at t_2.

The question of Maxwell's Demon is not only whether transitions that lead to anti-thermodynamic behavior are consistent with the (quantum or classical) dynamics, but whether there are cases in which such transitions are highly probable. We have just shown that there can be *no* no-go theorems in quantum mechanics which rule out a total decrease of entropy. And in particular, the quantum mechanical dynamics of measurement and erasure is compatible with such evolutions. We have no suggestion as to how to characterize in general thermodynamic entropy in quantum mechanics.

We know, for instance, that the von Neumann entropy does not in general behave like thermodynamic entropy. Therefore we deliberately

---

[20] This analysis applies *mutatis mutandis* to the GRW collapse theory, say with stochastic flashes. Here the stochastic dynamics of the flashes allows for a *micro*scopic erasure of memory.

formulated the question of Maxwell's Demon without appealing to the notion of *entropy*. However, assuming that one could come up with a quantum mechanical expression corresponding to thermodynamic entropy, we put forward the conjecture that in our microscopic set up the total entropy of the universe would decrease (with almost certainty) after each cycle.[21]

---

[21] The quantum mechanical uncertainty relations put constraints on the Demonic evolution described above. For example, inserting the partition is subject to the position-momentum uncertainty relations; likewise, the energy transfer from the environment to the particle is subject to the energy-time uncertainty relations. These effects would be significant for our argument if there had been a general theorem showing that they entail that more work must be invested in the process than can be produced by the Demon. We are not aware of such a theorem.

References

Albert, D. (1992) *Quantum Mechanics and Experience* Cambridge, MA: Harvard University Press.

Albert, D. (2000) *Time and Chance*. Cambridge, MA: Harvard University Press.

Albert, D. (2012). Physics and chance. In Y. Ben-Menahem and M. Hemmo (eds.), *Probability in Physics*. The Frontiers Collection, Berlin Heidelberg: Springer-Verlag, pp. 17–40.

Alexander, H. G (1956) *The Leibniz Clark Correspondence*. Manchester: Manchester University Press.

Arnold, V. I. and Avez, A. (1968) *Ergodic Problems of Classical Mechanics*. New York, Amsterdam: Benjamin.

Arntzenius, F. (2000) Are there really instantaneous velocities? *The Monist* **83**(2), 187–208.

Arntzenius, F. (2004) Time-reversal operations and the direction of time. *Studies in History and Philosophy of Modern Physics* **35**(1), 31–43.

Arntzenius, F. and Greaves, H. (2009) Time reversal in classical electromagnetism. *British Journal for the Philosophy of Science* **60**(3), 557–584.

Beisbart, C. and Hartmann, S. (eds.) (2011) *Probabilities in Physics*. Oxford: Oxford University Press.

Bell, J. S. (1987) Are there quantum jumps? In J. S. Bell (ed.), *Speakable and Unspeakable in Quantum Mechanics*. Cambridge: Cambridge University Press, pp. 201–212.

Ben Menahem (2011) The causal family: the place of causality in science. SSRN: http://ssrn.com/abstract = 1925493

Bennett, C. (1973) Logical reversibility of computation. *IBM Journal of Research and Development* **17**, 525–532.

Bennett, C. (1982) The thermodynamics of computation: a review. *International Journal of Theoretical Physics* **21**, 905–940.

Bennett, C. (2003) Notes on Landauer's principle, reversible computation, and Maxwell's Demon. *Studies in History and Philosophy of Modern Physics* **34**(3), 501–510.

Bergmann, P. G. and Lebowitz, J. L. (1955) New approach to nonequilibrium processes. *Physical Review* **99**, 578–587.

Berkovitz, J., Frigg, R. and Krontz, F. (2006) The ergodic hierarchy, randomness and chaos. *Studies in History and Philosophy of Modern Physics* **37**, 661–691.

Bernoulli, J. (1713) *The Art of Conjecturing, together with Letter to a Friend on Sets in Court Tennis.* English translation by Edith Sylla (2005). Baltimore: Johns Hopkins University Press.

Blatt, J. M. (1959) An alternative approach to the ergodic problem. *Progress of Theoretical Physics* **22**(6), 745–756.

Brading, K. and Castellani, E. (eds.) (2003) *Symmetries in Physics: Philosophical Reflections.* Cambridge: Cambridge University Press.

Brown, H. and Uffink, J. (2001) The origins of time-asymmetry in thermodynamics: the Minus First Law. *Studies in History and Philosophy of Modern Physics* **32**(4), 525–538.

Brown, H. Myrvold, W. and Uffink, J. (2009) Boltzmann's *H*-theorem, its limitations, and the birth of statistical mechanics. *Studies in History and Philosophy of Modern Physics* **40**, 174–191.

Brush, S. (1976) *The Kind of Motion We Call Heat.* Amsterdam: North Holland.

Bub, J. (2001) Maxwell's Demon and the thermodynamics of computation. *Studies in History and Philosophy of Modern Physics* **32**(4), 569–579.

Butterfield, J. (2011) Laws, causation and dynamics at different levels. PhilSci Archive, http://philsci-archive.pitt.edu/8745/

Butterfield, J. and Bouatta, N. (2011) Emergence and reduction combined in phase transitions. PhilSci Archive, http://philsci-archive.pitt.edu/8554/

Callender, C. (1995) The metaphysics of time-reversal: Hutchison on classical mechanics. *British Journal for the Philosophy of Science* **46**(3), 331–340.

Callender, C. (1999). Reducing thermodynamics to statistical mechanics: the case of entropy. *Journal of Philosophy* **XCVI**, 348–373.

Callender, C. (2001) Taking thermodynamics too seriously. *Studies in History and Philosophy of Modern Physics* **32**(4), 539–553.

Callender, C. (2004) Measures, explanation and the past: should "special" initial conditions be explained? *British Journal for the Philosophy of Science* **55**, 195–217.

Callender, C. (2007) The emergence and interpretation of probability in Bohmian mechanics. *Studies in the History and Philosophy of Modern Physics* **38**, 351–370.

Costa de Beauregard, O. and Tribus, M. (1974) Information theory and thermodynamics. *Helvetica Physica Acta*, **47**, 238–247.

Dürr, D., Goldstein, S. and Zanghi, N. (1992) Quantum equilibrium and the origin of absolute uncertainty. *Journal of Statistical Physics* **67**(5/6), 843–907.

Dürr, D. (2001) Bohmain mechanics. In J. Bricmont, D. Dürr, M. C. Galavotti *et al.* (eds.), *Chance in Physics: Foundations and Perspectives.* Lecture Notes in Physics, Springer-Verlag, pp. 115–132.

Earman, J. (1986) *A Primer on Determinism.* University of Western Ontario Series in Philosophy of Science, vol. 32, Dordrecht: Reidel.

Earman, J. (2002) What time-reversal invariance is and why it matters. *International Studies in the Philosophy of Science* **16**(3), 245–264.

Earman, J. (2006) The Past Hypothesis: not even false. *Studies in History and Philosophy of Modern Physics* **37**, 399–430.

Earman, J. and Norton, J. (1998) Exorcist XIV: the wrath of Maxwell's Demon. Part I. From Maxwell to Szilard. *Studies in History and Philosophy of Modern Physics* **29**(4), 435–471.

Earman, J. and Norton, J (1999) Exorcist XIV: the wrath of Maxwell's Demon. Part II. From Szilard to Landauer and beyond. *Studies in History and Philosophy of Modern Physics* **30**(1), 1–40.

Earman, J. and Redei, M. (1996) Why ergodic theory does not explain the success of equilibrium statistical mechanics. *The British Journal for the Philosophy of Science* **47**, 63–78.

Eddington, A. (1935) *The Nature of the Physical World*. London: Everyman's Library, J. M. Dent.

Ehrenfest, P. and Ehrenfest, T. (1912) *The Conceptual Foundations of the Statistical Approach in Mechanics*. Leipzig; reprinted 1990 New York: Dover.

Einstein, A. (1905) Über die von der molekularkinetischen theorie der wärme geforderte bewegung von in ruhenden flüssigkeiten suspendierten teilchen. *Annalen der Physik* **17**, 549–560. English translation in: R. Furth (ed.) Einstein, Albert, *Investigations on the Theory of Brownian Motion*, New York: Dover, 1926.

Einstein, A. (1970) Autobiographical notes. In P. A. Schilpp (ed.), *Albert Einstein: Philosopher-scientist*, vol. 2, Cambridge: Cambridge University Press.

Fahn, P. N. (1996) Maxwell's Demon and the entropy cost of information. *Foundations of Physics* **26**, 71–93.

Fermi, E. (1936) *Thermodynamics*. New York: Dover, 1956.

Feynman, R. (1963) *The Feynman Lectures on Physics*, Addison Wesley.

Feynman, R. (1965) *The Character of Physical Law*. Cambridge: MIT Press.

Feynman, R. (1996) *Feynman Lectures on Computation*. London: Penguin.

Fredkin, E. and T. Toffoli, T. (1982) Conservative logic. *International Journal of Theoretical Physics* **21**, 219–253.

French, A. P. and Taylor, E. F. (1978) *An Introduction to Quantum Physics*. Chapman & Hall.

Frigg, R. (2008) A field guide to recent work on the foundations of statistical mechanics. In D. Rickles (ed.), *The Ashgate Companion to Contemporary Philosophy of Physics*. London: Ashgate, pp. 99–196.

Frigg, R. and Werndl, C. (2012) A new approach to the approach to equilibrium. In Y. Ben-Menahem and M. Hemmo (eds.), *Probability in Physics*. The Frontiers Collection, Berlin Heidelberg: Springer-Verlag, pp. 99–114.

Ghirardi, G., Rimini, A. and Weber, T. (1986) Unified dynamics for microscopic and macroscopic systems. *Physical Review D* **34**, 470–479.

Gibbs, J. W. (1902) *Elementary Principles in Statistical Mechanics*. New Haven: Yale University Press.

Goldstein, S. (2001) Botlzmann's approach to statistical mechanics. In J. Bricmont, D. Dürr, M. C. Galavotti *et al.*, eds., *Chance in Physics: Foundations and Perspectives*. Springer-Verlag.

Goldstein, S. (2012). Typicality and notions of probability in physics. In Y. Ben-Menahem and M. Hemmo (eds.), *Probability in Physics*. The Frontiers Collection, Berlin Heidelberg: Springer-Verlag, pp. 59–72.

Guttmann, Y. (1999). *The Concept of Probability in Statistical Physics*. Cambridge: Cambridge University Press.

Hacking, I. (1975). *The Emergence of Probability*. Cambridge: Cambridge University Press.

Hahn, E. L. (1950). Spin echoes. *Physical Review* **80**, 580–594.

Hahn, E. L. (1953). Free nuclear induction. *Physics Today* **6**(11), 4–9.

Harman, P. M. (1982) *Energy, Force and Matter: The Conceptual Development of Nineteenth Century Physics*. Cambridge: Cambridge University Press.

Hawking, S. W. (1988) *A Brief History of Time*. London: Bantam Press.

Hemmo, M. and Pitowsky, I. (2007) Quantum probability and many worlds. *Studies in the History and Philosophy of Modern Physics* **38**, 333–350.

Hemmo, M. and Shenker, O. (2001) Can we explain thermodynamics by quantum decoherence? *Studies in the History and Philosophy of Modern Physics* **32**, 555–568.

Hemmo, M. and Shenker, O. (2003) Quantum decoherence and the approach to equilibrium (Part I). *Philosophy of Science* **70**, 330–358.

Hemmo, M. and Shenker, O. (2005) Quantum decoherence and the approach to equilibrium (Part II). *Studies in the History and Philosophy of Modern Physics* **36**, 626–648.

Hemmo, M. and Shenker, O. (2006) Von Neumann's entropy does not correspond to thermodynamic entropy. *Philosophy of Science* **73**(2), 153–174.

Hemmo, M. and Shenker, O. (2010) Maxwell's Demon. *The Journal of Philosophy* **107**, 389–411.

Hemmo and Shenker (2011a) Introduction to the philosophy of statistical mechanics: can probability explain the arrow of time in the Second Law of thermodynamics? *Philosophy Compass* **6/9**, 640–651.

Hemmo, M. and Shenker, O. (2011b) Szilard's perpetuum mobile. *Philosophy of Science* **78**, 264–283.

Hemmo, M. and Shenker, O. (2012) Measures over initial conditions. In Y. Ben-Menahem and M. Hemmo (eds.), *Probability in Physics*. The Frontiers Collection, Berlin Heidelberg: Springer-Verlag, pp. 87–98.

Hutchison, K. (1993) Is classical mechanics really time-reversible and deterministic? *British Journal for the Philosophy of Science* **44**, 341–347.

Jauch, J. M. and Baron, J. G. (1972) Entropy, information and Szilard's paradox. *Helvetica Physica Acta* **45**, 220–232.

Jaynes, E. T. (1957) Information theory and statistical mechanics. *Physical Review* **106**, 620–630 (Part I); *Physical Review* **108**, 171–190 (Part 2).

Jaynes, E. T. (1965) Gibbs vs. Boltzmann entropies. *American Journal of Physics*, **33**, 391. Reprinted in Jaynes, E. (1983), *Papers on Probability, Statistics and Statistical Physics*. Dordrecht: Reidel, pp. 79–88.

Knott, C. G. (1911) *Life and Scientific Work of Peter Guthrie Tait*. Cambridge: Cambridge University Press.

Kolmogorov, A. N. (1933) *Foundations of the Theory of Probability*, English translation 1956. Chelsea Publ. Co., New York.

Kripke, S. (1980) *Naming and Necessity*. Cambridge MA: Harvard University Press.

Lanczos, C. (1970) *The Variational Principles of Mechanics*. Toronto: University of Toronto Press.

Landau, L. D. and Lifshitz, E. M. (1980) *Statistical Physics Part 1, Course in Theoretical Physics* vol. **5**. 3rd ed. Translation by J. B. Sykes and M. J. Kearsley. Oxford: Butterworth-Heinemann.

Landauer, R. (1961) Irreversibility and heat generation in the computing process. *IBM Journal of Research and Development* **3**, 183–191.

Landauer, R. (1992) Information is physical. *Proceedings of PhysComp 1992, Workshop on Physics and Computation*. Los Alamitos: IEEE Computers Society Press, pp. 1–4.

Landauer, R. (1996) The physical nature of information. *Physics Letters A*, **217**, 188–193.

Lanford, O. E. (1975) Time evolution of large classical systems. In J. Moser (ed.), *Dynamical Systems, Theory and Applications*. Berlin: Springer, pp.1–111.

Lanford, O. E. (1976) On the derivation of the Boltzmann equation, *Asterisque* **40**, 117–137.

Lanford, O. E. (1981) The hard sphere gas in the Boltzmann–Grad limit. *Physica* **106A**, 70–76.

Lebowitz, J. (1993) Boltzmann's entropy and time's arrow. *Physics Today*, September, 32–38.

Leff, H. S. and Rex, A. (2003) *Maxwell's Demon 2: Entropy, Classical and Quantum Information, Computing*. Bristol: Institute of Physics Publishing.

Loewer, B. (2001) Determinism and chance. *Studies in History and Philosophy of Modern Physics* **32**, 609–620.

Malament, D. (2004) On the time reversal invariance of classical electromagnetic theory. *Studies in History and Philosophy of Modern Physics* **35B** (2), 295–315.

Maroney, O. (2005) The (absence of a) relationship between thermodynamic and logical reversibility. *Studies in History and Philosophy of Modern Physics* **36**, 355–374.

Maudlin, T. (2005) *The Metaphysics Within Physics*. Oxford: Oxford University Press.

Maudlin, T. (2007) What could be objective about probabilities? *Studies in History and Philosophy of Modern Physics* **38**, 275–291.

Maxwell, J. C. (1868) Letter to W. G. Tait. In C. G. Knott, *Life and Scientific Work of William Guthrie Tait*. London: Cambridge University Press, 1911.

North, J. (2011) Time in thermodynamics. In C. Callender (ed.), *The Oxford Handbook of Philosophy of Time*. Oxford: Oxford University Press, pp. 312–350.

Penrose, O. (1970) *Foundations of Statistical Mechanics*. New York: Pergamon Press.

Peres, A. (1993) *Quantum Theory: Concepts and Methods*. Dordrecht: Kluwer.

Pitowsky, I. (1992) Why does physics need mathematics? A comment. In E. Ulmann-Margalit (ed.), *The Scientific Enterprise*, Dordrecht, Kluwer, pp. 163–167.

Pitowsky, I. (2012) Typicality and the role of the Lebesgue measure in statistical mechanics. In Y. Ben-Menahem and M. Hemmo (eds.), *Probability in Physics*. The Frontiers Collection, Berlin Heidelberg: Springer-Verlag, pp. 41–58.

Price, H. (1996) *Time's Arrow and Archimedes' Point: New Directions for the Physics of Time*. New York: Oxford University Press.

Putnam, H. (1967) Psychological predicates. In W. H. Capital and D. D. Merrill (eds.), *Art, Mind and Religion*. Pittsburgh: University of Pittsburgh Press; reprinted and retitled: The nature of mental states. In H. Putnam, *Mind, Language, and Reality, Philosophical Papers*, vol. 2, 2nd edition. Cambridge: Cambridge University Press (1979).

Reichenbach, H. (1956) *The Direction of Time*. Berkeley: University of California Press.

Ridderbos, K. (2002) The coarse graining approach to statistical mechanics: how blissful is our ignorance? *Studies in History and Philosophy of Modern Physics* **33**(1), 65–77.

Ridderbos, K. and Redhead, M. (1998) The spin echo experiments and the Second Law of thermodynamics. *Foundations of Physics* **28**(8), 1237–1270.

Russell, B. (1912) On the notion of cause. *Proceedings of the Aristotelian Society* New Series **13**, 1–26.

Russell, B. (1921) *The Analysis of Mind*, Lecture IX.

Shalizi, C. R. and Moore, C. (2003) What is a macrostate? Subjective observations and objective dynamics. arXiv:cond-mat/0303625v1 [cond-mat.stat-mech].

Shenker, O. (1999) Maxwell's Demon and Baron Munchausen: free will as a *perpetuum mobile*. *Studies in the History and Philosophy of Modern Physics* **30**, 347–372.

Sklar, L. (1973) Statistical explanation and ergodic theory. *Philosophy of Science* **40**, 194–212.

Sklar, L. (1993) *Physics and Chance*. Cambridge: Cambridge University Press.

Sklar, L. (2000) *Theory and Truth* (Oxford: Oxford University Press).

Szilard, L. (1929). On the decrease in entropy in a thermodynamic system by the intervention of intelligent beings. In: H. S. Leff and A. Rex, A. (eds.), *Maxwell's Demon 2: Entropy, Classical and Quantum Inforamtion, Computing*. Bristol: Institute of Physics Publishing, 2003, pp. 110–119.

Swendsen, R. H. (2011) How physicists disagree on the meaning of entropy. *American Journal of Physics* **79**, 342–348.

Tolman, R. (1938) *The Principles of Statistical Mechanics*. New York: Dover, 1979.

Tribus, M. and McIrvine, E.C. (1971) Energy and information. *Scientific American* **224**, 179–186.

Tumulka, R. (2006) A relativistic version of the Ghirardi–Rimini–Weber Model. *Journal of Statistical Physics* **125**, 821–840.

Uffink, J. (2001) Bluff your way in the Second Law of thermodynamics. *Studies in History and Philosophy of Modern Physics* **32**, 305–394.

Uffink, J. (2006) Compendium to the foundations of classical statistical physics. In J. Butterfield and J. Earman (eds.), *Handbook for the Philosophy of Physics*, Part B, pp. 923–1074.

Uffink, J. (2008) Boltzmann's work in statistical physics. In E. N. Zalta (ed.) *The Stanford Encyclopedia of Philosophy* (Winter 2008 Edition), http://plato.stanford.edu/entries/statphys-Boltzmann.

Uffink, J. and Valente, G. (2010) Time's arrow and Lanford's theorem. *Seminaire Poincaré XV Le Temps*, 141–173.

van Fraassen, B. (1989) *Laws and Symmetry*. Oxford: Clarendon Press.

von Neumann, J. (1932) *Mathematical Foundations of Quantum Mechanics*, English translation by R. T. Beyer (1955). Princeton: Princeton University Press.

von Plato, J. (1994) *Creating Modern Probability*. Cambridge: Cambridge University Press.

Walker, G. H. and Ford, J. (1969) Amplitude instability and ergodic behaviour for conservative nonlinear oscillator Systems. *Physical Review* **188**, 416–32.

Winsberg, E. (2004). Can conditioning on the "Past Hypothesis" militate against the reversibility objections? *Philosophy of Science* **71**, 489–504.

Zurek, W. (1990). Algorithmic information content, Chuch–Turing thesis, physical entropy and Maxwell's Demon. In W. Zurek (ed.), *Complexity, Entropy and the Physics of Information*. Redwood: Addison Wesley.

Index